SpringerBriefs in Applied Sciences and Technology

PoliMI SpringerBriefs

Springer, in cooperation with Politecnico di Milano, publishes the PoliMI SpringerBriefs, concise summaries of cutting-edge research and practical applications across a wide spectrum of fields. Featuring compact volumes of 50 to 125 (150 as a maximum) pages, the series covers a range of contents from professional to academic in the following research areas carried out at Politecnico:

- Aerospace Engineering
- Bioengineering
- Electrical Engineering
- Energy and Nuclear Science and Technology
- Environmental and Infrastructure Engineering
- Industrial Chemistry and Chemical Engineering
- Information Technology
- Management, Economics and Industrial Engineering
- Materials Engineering
- Mathematical Models and Methods in Engineering
- Mechanical Engineering
- Structural Seismic and Geotechnical Engineering
- Built Environment and Construction Engineering
- Physics
- Design and Technologies
- Urban Planning, Design, and Policy

http://www.polimi.it

Simona Chiodo · David Kaiser · Julie Shah ·
Paolo Volonté
Editors

Improving Technology Through Ethics

POLITECNICO
MILANO 1863

Springer

Editors
Simona Chiodo
Department of Architecture and Urban Studies
Politecnico di Milano
Milan, Italy

Julie Shah
Department of Aeronautics and Astronautics
Massachusetts Institute of Technology
Cambridge, MA, USA

David Kaiser
Program in Science, Technology, and Society
Department of Physics
Massachusetts Institute of Technology
Cambridge, MA, USA

Paolo Volonté
Department of Design
Politecnico di Milano
Milan, Italy

ISSN 2191-530X ISSN 2191-5318 (electronic)
SpringerBriefs in Applied Sciences and Technology
ISSN 2282-2577 ISSN 2282-2585 (electronic)
PoliMI SpringerBriefs
ISBN 978-3-031-52961-0 ISBN 978-3-031-52962-7 (eBook)
https://doi.org/10.1007/978-3-031-52962-7

This Springer imprint is published by the registered company Springer Nature Switzerland AG
The registered company address is: Gewerbestrasse 11, 6330 Cham, Switzerland

Paper in this product is recyclable.

Contents

What Improving Technology Through Ethics Means

Simona Chiodo, David Kaiser, Julie Shah, and Paolo Volonté

1 Technology and Ethics

In a 1996 conference in Italy, Kenneth Keniston, one of the founders of the Program in Science, Technology and Society at the Massachusetts Institute of Technology (MIT), diagnosed a crisis of the engineering algorithm [1]. By "engineering algorithm" he meant the traditional epistemic culture of Western engineering that is based on the separation of two realms of meaning: the realm of problems that can be solved in principle through the use of technology, mathematics and scientific knowledge; and the realm of everything else, including, among others, social problems, value issues and philosophical or religious matters. In short, the engineering algorithm holds that, in order to effectively address the first type of problems, technologists must be completely disinterested in the second type of problems. To be a good engineer, you should disregard any social issues and delegate ethical concerns to others.

According to Keniston, the engineering algorithm was already in a state of crisis at the time of his talk. Among the main reasons for this crisis, he identified the increasing complexity of emerging technologies. The lone engineer designing a single product had been replaced by the coordinated team of interacting specialists working on the complex design of a specific component of an elaborate socio-technical system. The consideration of trade-offs between incommensurable factors—such as efficiency

S. Chiodo (✉)
Department of Architecture and Urban Studies, Politecnico di Milano, Milan, Italy
e-mail: simona.chiodo@polimi.it

D. Kaiser
Department of Physics, Massachusetts Institute of Technology, Cambridge, MA, USA

J. Shah
Department of Aeronautics and Astronautics, Massachusetts Institute of Technology, Cambridge, MA, USA

P. Volonté
Department of Design, Politecnico di Milano, Milan, Italy

S. Chiodo et al. (eds.), *Improving Technology Through Ethics*,
PoliMI SpringerBriefs, https://doi.org/10.1007/978-3-031-52962-7_1

and safety—was moving towards the centre of engineering. And what used to be dealt with as an externality, such as environmental impact, was increasingly becoming an integral part of engineering design. To use his words, "'society', once something out there, had entered the workplace" [1].

Almost thirty years have passed since Keniston diagnosed this crisis, but things have not changed. Quite the opposite, the advancing digitization of society has underscored the relevance of ethical concerns in the development and use of technology. The pervasive nature of digital technologies in the contemporary world, that is, their presence in and relevance to almost all actions in the everyday life of ordinary people, has caused them to become one of the main sources of concern for issues such as privacy, security, safety, sustainability, justice, inequality, and discrimination. Widely debated are cases such as the power of algorithms and AI to incorporate and reproduce traditional social prejudices in the form of coded bias [2] or how digital twins can worsen one's present by predicting one's future in detail, to the point that "[r]esearchers report having identified biomarkers that help in predicting when a person will die and from what cause" [3].

However, this hendiadys of technology and ethics is not novel. In recent decades, humanists and social scientists have frequently argued that the social impact of technological innovation requires more and more ethical awareness (see at least [4–17]). It is therefore no coincidence that the number of journals and book series that focus on the relationship between technology and ethics has exponentially increased in the humanities (see, for instance, the seminal journal "Ethics and information technology" founded in 1999) and social sciences (see, for instance, the seminal journal "Science, technology and human values" founded in 1976).

Even though authors explicitly encourage dialogue between the humanities and the social sciences, on the one hand, and science and engineering, on the other hand, those who publish in such journals are usually humanists and social scientists, and coauthorships with scientists and engineers remain uncommon. For instance, although the last issue of "Ethics and information technology"[1] features articles written either by human and social scientists or by scientists and engineers, only one article out of eleven is coauthored by scholars from both subject areas. Similarly, in the last issue of "Science, technology and human values",[2] where the number of social scientists is definitely higher than the number of other scholars, only one article out of nine is coauthored by scholars from both subject areas.

It seems that our new awareness about the "moral status" of technologies [17] is not particularly effective for encouraging philosophers and engineers to collaborate. There are several obstacles to an authentic dialogue. First, there are scientific reasons: research methodologies are so different to the point that, even though the basic vocabulary of both subject areas may share similar words, their meanings may be completely different. For instance, while for a moral philosopher "autonomy" means the ability of the moral subject, typically a human being, to choose freely and autonomously without being coerced, for a technologist "autonomy" means the

[1] See https://link.springer.com/journal/10676/volumes-and-issues/24-4, last accessed 2023/1/3.

[2] See https://journals.sagepub.com/toc/STH/current, last accessed 2023/1/3.

ability of a technological device to operate without having to wait for instructions or interventions from a human being. Second, there are research policy reasons: interdisciplinarity and transdisciplinarity are encouraged in theory (for instance, by establishing national and supranational guidelines for rewarding such research), but they are discouraged in practice (for instance, because recruitment policies devalue them, and authors focused on their specific disciplines are more likely to get hired than colleagues who coauthor articles with scholars from other disciplines). Third, there are institutional reasons: the organization of most universities into separate departments encourages scholars to spend time with colleagues from the same discipline, and therefore causes it to be more likely that they will collaborate with each other. Even ethicists who work at one of the world's most pioneering institution in terms of integrating the ethics of technology, namely TU Delft, underscore the limits resulting from the division described above: they admit they are more likely to coauthor publications on technology ethics with colleagues from their department than with scientists and engineers from other departments.[3]

The end result is that, even though the hendiadys of technology and ethics not only exists as a concept but is also increasingly deployed, its effectiveness is limited by pre-existing boundaries between academic disciplines: those who can make a real difference towards achieving the ethical development of technology, i.e. scientists and engineers, usually read what other scientists and engineers publish rather than what humanists and social scientists, and especially ethicists, write. Thus, even though the latter group critically reflects upon the ethics of technology, the former does not usually embed such reflections into the design of technology. Conversely, humanists and social scientists often raise ethical questions when it is too late to make a real difference, i.e. when scientists and engineers have already designed technologies that may cause disruptive effects. Moreover, the potential effectiveness of the hendiadys of technology and ethics is further limited when human and social scientists develop an understanding of technology that is substantially disconnected not only from scientists' and engineers' practices or processes of technology design, but also from issues that arise from the distribution and use of technology.

2 Technology Through Ethics

The hendiadys of technology and ethics is therefore not a suitable concept for describing what is needed to cope with technology's present and future societal challenges. We do not just need more ethics. We need a qualitative shift. In this volume, we propose that rather than delegating the task of evaluating technological outcomes to moral philosophers and social scientists, we need to make the technological system capable of designing technology through ethics.

[3] This results from a survey carried out at Politecnico di Milano in 2017–18: "Rapporto sulla presenza e integrazione delle discipline META nei politecnici e nelle università tecniche in Europa", unpublished draft, June 2018.

The volume collects contributions by junior scholars at the Massachusetts Institute of Technology and Politecnico di Milano to illustrate various approaches to the design of technology through ethics; at the same time, it is also a testimony to the collaboration between the two universities towards tackling this challenge. In what follows, we shall discuss some crucial aspects of this approach.

The education of future scientists and engineers—those who can make a real difference—is a key issue here. First, including ethics and other subjects aimed at cultivating critical thinking in STEM curricula would help. This is already happening at several European and North-American universities. At TU Delft, for instance, the "Ethics and Philosophy of Technology Section" offers training in ethics to all engineering students, from courses focused on specific technologies to more interdisciplinary courses.[4] Similar plans are being implemented at RWTH Aachen, Politecnico di Torino, École Polytechnique Fédérale de Lausanne, and in several other technical universities in Europe. Similarly, the cross-disciplinary initiative in Social and Ethical Responsibilities of Computing (SERC) at MIT has gathered humanists and social scientists together with scientists and engineers to create new pedagogical materials and to share them broadly via open-access publishing.[5] Second, better teaching outcomes are achieved by integrating both kinds of expertise, which means that courses on technology ethics should be taught by philosophers and engineers side by side. The integration of these two kinds of expertise may be a promising way to preserve them: not only science and engineering but also ethics, as an almost three-millennia-old branch of philosophy, requires specialist expertise not to fall into amateurism. Thus, in courses of technology ethics, scientists and engineers are needed both to identify concrete technological issues in detail and to translate the proposals put forward by humanists and social scientists into concrete solutions. Humanists and social scientists are, in turn, needed to offer their expertise both on societal problems (from historical developments to present and future social impacts) and additional forms of reasoning (conceptual and qualitative analysis, philosophical abstraction and idealization, and thought experiments).

We ourselves gained experience with the integration of these two kinds of expertise when we selected Ph.D. students and early-career researchers from our universities and trained them intensively in technology ethics. Interdisciplinarity was crucial for us not only for the reasons described above but also because it is a key issue when it comes to moving from theory to practice: Ph.D. students from leading technical universities are likely to contribute to future decisions that will involve the assessment of ethical issues and societal challenges raised by emerging technologies; creating a common ground now will facilitate mutual understanding once they will hold positions of responsibility.

---

[4] The list is available at https://www.tudelft.nl/ethics/ethics/teaching-activities/ethics-teaching, last accessed 2023/1/10.

[5] These resources include the peer-reviewed MIT SERC Case Studies series available at https://mit-serc.pubpub.org (last accessed 2023/4/11) and the SERC classroom resources available at https://ocw.mit.edu/courses/res-tll-008-social-and-ethical-responsibilities-of-computing-serc/ (last accessed 2023/4/11).

Research should also pursue the integration of these two kinds of expertise. Several technical universities have realized that having departments that are organized according to disciplinary boundaries can limit effective integration and favor inflexibility and narrow-mindedness. Within this approach, a new initiative launched by Politecnico di Milano is the integration of philosophers and social scientists in existing engineering departments—thus avoiding the creation of a new department or research center devoted to the humanities and social sciences and introducing instead ethical, sociological, philosophical, historical and political-science perspectives into environments that are dedicated to technological research. Sharing physical space in this manner, as we partly started experiencing, facilitates the integration of different kinds of expertise from daily conversations in front of the coffee vending machine to planned research initiatives, starting with writing research projects together and co-authoring publications.

Ethics committees may be a further way to pursue integration, as they force members of such committees to hear each other's perspectives on the ethical relevance of research processes and technical solutions. As we partly started experiencing, when a computer scientist, a computer engineer, a philosopher of technology and a sociologist of technology lock themselves in a room to try to answer a moral dilemma raised by a concrete research project, they cannot simply give up on the discussion when achieving mutual understanding is difficult. Instead, they have to learn to join forces for the sake of tackling a real-world problem that otherwise risks having a dramatic impact on humans. Similarly, according to Michelle Lamont, deferring to the expertise of others is considered a foundational rule for interdisciplinary scientific research evaluation panels, and being able to articulate one's point of view to others outside one's field is considered a crucial skill for being a good panelist [18].

An authentic dialogue between the humanities and social sciences, on the one hand, and science and engineering, on the other hand, is also hampered by the currently prevailing hierarchy of values in the Global North, and by the manner in which mainstream academia currently privileges productivity and performance over doubt and critical thinking. The result is that, whenever the humanities and social sciences risk slowing down technological research by raising annoying moral dilemmas, it becomes more likely that humanists and social scientists are cordoned off into their own departments, where they subsequently run the risk of falling back into their worst habits, such as employing "ethics" as an empty buzzword without having any relevant knowledge about technological design and usage. Questioning the predominant hierarchy of values is thus a crucial condition, for scientists and engineers, to be happy to slow down for the sake of the common good.

Values are difficult to change, especially if they are the expression of a culture that, as in the case of Western modernity, is rooted in the process of individualization [19] against the pre-modern dominance of community. Yet, values can change, and even the dominating hierarchy of values can change surprisingly quickly. For instance, in Western cultures the women's ideal of labor has moved in a few decades from carrying out unpaid care work that is dedicated to home and family to the construction of professional careers and economic independence in what has been called a "quiet revolution" [20]. Hence, it is not necessary to ask for a revolution. Rather, we urge a quiet evolution in the understanding of values, moving from ethics as a mere rhetoric tool to ethics as a driver of development.

In order to explore this, we invited early-career researchers at MIT and Politecnico di Milano, who were intensively trained in technology ethics through a joint program, to contribute to this volume with the following specific goal: to show what improving technology through ethics can really mean. As we shall see in the next section, they responded to our invitation by proposing a variety of issues upon which to reflect: present difficulties in making the collaboration between ethics and technical disciplines effective; how to overcome such obstacles right now and in the future; what scientists and engineers can gain from working with ethicists side by side; and what the latter can gain from the former, specifically when moving from abstract theoretical ethics to co-designing with scientists and engineers technologies for the common good. That these promising early-career researchers have been provided with a new awareness about technology ethics is our hope for a better future.

3 Cases in Point

Most interestingly, the authors of the following chapters generally embody a growing phenomenon in the academic community: their research profiles often combine expertise coming from science and engineering, on the one hand, with expertise coming from the humanities and the social sciences, on the other hand. This does not mean, and should not mean, that they have lost the kind of "vertical" perspective that makes researchers real experts in their fields. Conversely, it might mean that, as complexity and uncertainty increasingly characterize our age, adding a kind of "horizontal", i.e. forward-looking, perspective may be promising, and even wise. For instance, it would be senseless, not to say dangerous, to develop technologies that seek to increase the life expectancy of humans within a few generations without considering, by means of a forward-looking perspective, how to face the impact of such technologies in terms of natural resources, energy justice, social justice, the meaning of human life, and so forth. Thus, through the following chapters, we are happy to demonstrate through a variety of ways that adding "horizontality" to an already well-established "verticality" may be the best and most responsible way to carry out technical research.

The first chapter ("An Ethnographer Among the Engineers: Doing STS at a Technical University"), by Chris Hesselbein, is an illuminating meta-reflection upon the present difficulties in making the collaboration between ethics and technical disciplines effective, as well as how it is possible to overcome these obstacles. Specifically, Chris Hesselbein provides a personal reflection upon the experience of being an integrated social scientist at a technical university, both when it comes to teaching sociological perspectives on technoscience to engineering students and when it comes to doing research with engineering colleagues. As we have already mentioned in the previous sections, the challenges to address are several, from being interpersonal to being practical, and may also include tensions at the institutional level. Yet, critical awareness of difficulties is the first step to try to overcome them, starting with improving the professional role of integrated social scientists themselves and, thus, trying to provide the next generations of both social scientists and engineering colleagues and students with improvements of the current state of affairs.

In the second chapter ("Future Medicine: Towards a More Conscious and Ethical Communication"), Beatrice Lucia Bona, as a scientist working on nanomedicine, shows how a better ethical awareness can improve scientific communication as an important part of scientists' responsibility. The recent development of extremely innovative diagnostic and therapeutic technologies gives hope to patients suffering from pathologies for which today there is still no cure, starting with nanotherapies that have just reached the clinical application and, thus, have just moved from the domain of a restricted group of experts to the domain of common people. This move also implies ethical and social concerns, from how transparent and ethical professionals' decisions are to how nanotechnology research and applications are presented to citizens, especially when it comes to encouraging public trust. Thus, Beatrice Lucia Bona stresses the importance of learning how to communicate medical innovations through a new awareness of scientists' ethical and social role, especially in an age characterized by both great innovation and immediate spread of news, sometimes uncritically.

The third chapter ("Data Quality, Data Diversity and Data Provenance from an Ethical Perspective"), by Edoardo Ramalli, is also written from a scientific perspective. As a computer scientist, Edoardo Ramalli explores some kinds of ethical issues involved in predictive models, which, in the last years, have been remarkably strengthened by the increasing amount of data we have at our disposal. After a phase in which accuracy and efficiency were the main issues to be addressed in the development of predictive models, the focus has recently shifted from what can be done with data to what we should or should not do with it. In this sense, the author takes into account data quality as one of the main ways to address predictive models' ethical issues. Also, other data-related aspects are relevant when it comes to predictive models' impacts, from data diversity to data provenance, which should be regularly addressed by the data science debate. This chapter contributes to this debate through a bottom-up approach, namely, by analyzing ethical issues arising from data quality, diversity and provenance, together with their possible trade-offs and mitigations.

Both the fourth chapter and the fifth chapter are connected to the realm of technologies applied to the city. In the first case ("Ethics-aware Application of Digital Technologies in the Construction Industry"), Ania Khodabakhshian reflects upon the relationship between technological innovation and the construction industry as a sector usually suffering from inefficient decisions, time-consuming activities, unproductivity, resistance to change, and high rates of accidents. Thanks to the recent advancements of technologies such as artificial intelligence and digital twins, the construction industry is experiencing a drastic shift toward digitization, automation, and optimization, which may offer the perfect solution to what mentioned above. But emerging technologies' potential harms are still overlooked. Ania Khodabakhshian focuses on the development of ethical standards and regulations in the design of fairer systems, starting with the idea of assessing productivity not only financially but also socially. After having stressed the key role played by designers when it comes to (inevitably) embedding ethical and social values or disvalues in their artifacts, the author examines the ethical issues that may arise from emerging technologies applied to the construction industry in order to define possible criteria as drivers of an ethics-aware technology implementation framework.

In the second case ("Drawing Together: Technology Development for Public Service"), Carlos Sandoval Olascoaga, Lillian Chin, Josefina Correa Menendez, Reid Kovacs, Calvin Zhong and Nicholas de Monchaux show how a participatory software design and workforce development project can help tackle the problems affecting the New York City Housing Authority residents, such as systemic barriers to economic mobility, pervasive social inequities, poverty, and racial discrimination. Thanks to a collaboration with the Green City Force, which developed a multi-site and multi-partner initiative to improve residents' health and wellbeing through community spaces combining community engagement, workforce development, healthy food production, and sustainable landscape transformations to existing infrastructure, the authors explore the ways in which computing can be leveraged to co-design a large number of network-based urban interventions and to measure the ecological and physical impact of these proposals in the overall urban resilience. Thus, the community-driven software development process is analyzed, which aims to ensure community participation and co-design not only in terms of software infrastructures but also in terms of physical spaces. Also, an educational framework for incorporating community-based technology development into workforce development curricula is described.

The social aspects of emerging technologies are also key in the sixth chapter ("Addressing Big Data Poverty in Sub-Saharan Africa for Responsible AI Development"), in which Leonard Francis Vibbi highlights the critical issues of data poverty and lack of data regulations in Sub-Saharan Africa, as well as their impacts on artificial intelligence (AI) responsible development. The author defines big data poverty as a scarcity of high volume, velocity, and variety of information that can be used for high-quality training of machine learning algorithms to develop and deploy AI technologies. By taking into account the case of the COVID-19 pandemic, the author discusses the impacts of data poverty and lack of data regulations on individual privacy, data protection, and responsible data use in AI development, which can easily lead not only to the misuse of personal data but also to undermining public trust in AI technologies. Therefore, as Leonard Francis Vibbi argues, policymakers and stakeholders in Sub-Saharan Africa should prioritize the best practices to enhance the region's potential for AI development, as well as to ensure that these technologies are utilized responsibly and ethically.

The seventh chapter ("Social and Technological Innovation: Cross-Fertilization Needed") takes a socio-economic perspective to focus, again, on the social aspects of emerging technologies by somehow turning the tables: the former can especially benefit from the latter. But, as Federico Bartolomucci, Giorgia Trasciani and Francesco Gerli argue, ethical awareness is key in this respect, in that technological applications aiming to tackle societal challenges frequently risk having unintended negative consequences, particularly for underrepresented and disadvantaged groups.

This is the case with data-intensive technologies and applications for the sharing economy, for instance. Combining the conventional tech innovation process with a few social innovation distilled features can be a promising way to prevent the creation of negative externalities. On the one hand, it can be advantageous for both technological and social actors. On the other hand, it can strengthen the collective capacity to generate impact through technology. Federico Bartolomucci, Giorgia Trasciani and Francesco Gerli show that cross-fertilization requires to rethink the innovation processes and the roles that various actors play inside it, as well as to reconceptualize innovation drivers and value redistribution mechanisms. Moreover, the authors identify two arenas where cross-fertilization could occur: first, the management and development of innovation processes and, second, the competencies involved. Finally, by closing the loop of the first chapter, the authors conclude with a reflection upon the role that universities, from research to teaching to public engagement, can play in this scenario.

References

1. Keniston K (1996) The crisis of the engineering algorithm. Talk for the Institute for Advanced Studies in the Humanities, Politecnico di Torino, 17 Oct 1996
2. O'Neil C (2016) Weapons of math destruction: how big data increases inequality and threatens democracy. Crown, New York
3. Gigerenzer G, Garcia-Retamero R (2017) Cassandra's regret. The psychology of not wanting to know. Psychol Rev 2(124):179–196
4. Winner L (1977) Autonomous technology: technics-out-of-control as a theme in political thought. MIT Press, Cambridge-London
5. Ihde D (1979) Technics and praxis. Reidel, Dordrecht-Boston-Lancaster
6. Baum RJ (1980) Ethics and engineering curricula. The Hastings Center, Hastings-on-Hudson
7. Borgmann A (1984) Technology and the character of contemporary life. University of Chicago Press, Chicago
8. Jonas H (1984) The imperative of responsibility: in search of an ethics for the technological age. University of Chicago Press, Chicago-London
9. Latour B (1993) We have never been modern. Harvester Wheatsheaf, New York
10. Harris CE, Pritchard MS, Rabins M (1994) Engineering ethics: concepts and cases. Wadsworth, Belmont
11. Davis M (2005) Engineering ethics. Ashgate, Aldershot-Burlington
12. Wallach W, Allen C (2009) Moral machines: teaching robots right from wrong. Oxford University Press, Oxford-New York
13. Floridi L (ed) (2010) The Cambridge handbook of information and computer ethics. Cambridge University Press, Cambridge
14. van de Poel I, Royakkers L (2011) Ethics, technology and engineering. Wiley-Blackwell, Oxford
15. Verbeek PP (2011) Moralizing technology: understanding and designing the morality of things. The University of Chicago Press, Chicago-London
16. van den Hoven J, Vermaas PE, van de Poel I (eds) (2015) Handbook of ethics and values in technological design: sources, theory, values and application domains. Springer, Dordrecht
17. Kroes P, Verbeek PP (eds) (2014) The moral status of technical artefacts. Springer, Dordrecht
18. Lamont M (2009) How professors think: inside the curious world of academic judgment. Harvard University Press, Cambridge-London

19. Beck U, Beck-Gernsheim E (2002) Individualization: institutionalized individualism and its social and political consequences. Sage, London
20. Goldin C (2006) The quiet revolution that transformed women's employment, education and family. Am Econ Rev 96(2):1–21

An Ethnographer Among the Engineers: Doing STS at a Technical University

Chris Hesselbein

1 Introduction

The increasingly widespread realization that innovations in science and technology do not automatically lead to socio-economic or environmental progress—and indeed can have unforeseen, negative consequences and lead to public controversy—has given rise to initiatives that seek to complement dominant technoscientific narratives about research, development, and public policy with socially-informed perspectives from the humanities and social sciences. These initiatives range from "responsible research and innovations" (RRI) programs to public engagement exercises that strive to bridge the gap between science and society by involving citizens in decisions about science and technology to better align the interests of society with the practices and outcomes of technoscience. In addition, efforts are also being made to bring social and ethical dimensions back into the spaces and practices of scientists and engineers, thus improving their awareness of the societal relevance and impact of their work.

One currently popular variation of these initiatives is the "integration" of scholars from the social sciences and humanities in technology companies, research centers, and technical universities, where they are expected to "critically participate" in research, development, and governance processes in order to maximize social benefits and reduce the possibility of negative impacts and public controversy. In public research institutions and technical universities in particular, integrated social scientists are expected to teach science and engineering students about the social and ethical dimensions of their work as well as collaborate with engineering colleagues in an interdisciplinary manner. This is a valuable opportunity for social scientists not only to make engineering students as well as professional engineers more critically aware about social concerns and the role that technoscientific practices and products

C. Hesselbein (✉)
Department of Management, Economic and Industrial Engineering, Politecnico di Milano, Milan, Italy
e-mail: christopher.hesselbein@polimi.it

S. Chiodo et al. (eds.), *Improving Technology Through Ethics*,
PoliMI SpringerBriefs, https://doi.org/10.1007/978-3-031-52962-7_2

play in society, but also to underscore the responsibility of engineers as technical experts with a great degree of authority and to highlight the social, cultural, and political implications of their work.

Interacting as well as collaborating with engineers—and therefore potentially exerting a direct influence on the training and professional practices of engineers—is an uncommon privilege for scholars of technoscience in the social sciences and humanities. However, relatively little research has been done on the influence or impact that integrated scholars have on the engineering profession and the challenges they face as interdisciplinary and "integrated" scholars. The few publications that do exist consist mainly of first-person, personal accounts from social scientists rather than systematic analyses [1–4]. We still know little, for example, about how scientists and engineers perceive the role of social scientists in their departments and the contributions they make (or problems they cause) as well as the incentives and barriers for engaging in interdisciplinary research [5, 6].

Moreover, the valuable opportunity to intervene in the work of scientists and engineers can lead to various professional challenges for social scientists who are expected to fulfil such a role. The critiques from scholars of technoscience tend to be made from a safe distance, ensconced in home departments, and delivered to core journals that are not often read by scientists and engineers. As "integrated" scholars, however, we are necessitated to deliver our critiques in a more direct and face-to-face manner, required to practically collaborate with our engineering colleagues, and remain financially dependent on our departmental and institutional directors. In other words, despite the laudable ideas and valuable opportunities that underpin our "integration" in engineering departments, this endeavor can also result in lingering tensions and frustrations that can adversely affect not only our careers but also our ability to deliver the perspective and critique that the social sciences and humanities are purportedly expected to provide.

In this chapter, I reflect on my own position and experience as one such "integrated" social scientist at a major polytechnical university in northern Italy, namely the Politecnico di Milano. I first discuss my role as an instructor of a course on the sociology of technology as well as a co-instructor of an "integrated course" on "diversity-aware design", which is co-taught with an engineering colleague. Here, I reflect on the various issues and negotiations that emerge in teaching for and with engineers as well as on some of the pedagogical approaches that have proven most useful for introducing critical concepts from Science and Technology Studies (STS). Second, I discuss my role as a social science researcher who is expected to collaborate with engineering colleagues, and I reflect on the tensions that arise from differences in disciplinary frameworks, theoretical languages, epistemic goals and methods as well as diverging professional or institutional imperatives and demands. In conclusion, I make three very brief suggestions for further strengthening the role of the social sciences for both teaching and collaborating in engineering spaces.

Although this chapter is written in the first person, and by necessity a personal reflection, I do not wish to represent myself as either being more "virtuous" than my engineering colleagues or as having "privileged knowledge" that they do not have [7]. I merely wish to reflect on and recount my experiences as someone who is trained in STS and qualitative methods, and who has, at the time of this writing, accumulated over two years of experience as an integrated social scientist teaching engineering students and working with quantitative researchers and engineers. Placing myself at the center of this story is, moreover, an uncommon exercise that my training has generally warned me against even if reflexivity and the acknowledgement of one's identity as a researcher are central tenets of STS and the more ethnographically-oriented disciplines. By describing my personal experiences, I hope to give a partial and personal perspective on the challenges, but also pleasures, that may emerge from being an integrated social scientist at a technical university. The use of the first-person perspective should not, however, be taken as an indication that the practices and problems I describe are merely individual. Rather, these are collective and systemic challenges that are becoming more and more relevant, not least because the integration of scholars from the social sciences and humanities in the study and management of technoscience is increasingly going "mainstream" even if its value has not yet been adequately examined [2, 8].

2 Institutional Integration of Non-engineering Perspectives

The Politecnico di Milano—the largest technical university in Italy, which was founded in 1863 and is considered a global leader in multiple engineering disciplines—formally initiated the institutional integration of the social sciences and humanities in 2016 with the founding of a "study unit" called META. Under the banner of "social sciences and humanities for science and technology", META consists of an interdisciplinary network of (un-)tenured professors, postdocs, and PhD students with expertise in the epistemological, ethical, and social implications of technological and scientific research and development. At the moment, the membership of META (a total of 23 scholars) is predominantly made up of philosophers of science, with the remainder consisting of—in descending numbers—sociologists, other kinds of philosophers, historians, media scholars, anthropologists, and literary scholars.

Because META is not a formal department or program, and therefore not allocated its own physical space within the university, its members are all based in separate departments of the university, such as Design, Urban Studies and Architecture, and the various subdisciplines of Engineering. In other words, although META is formally set up to provide a critical outsider perspective, its members are not physically or organizationally segregated from the engineers and designers but instead deeply entrenched within their various departments. This, as I will discuss in more detail below, forces META members to take on a double identity and role, which can have significant benefits but also drawbacks.

META offers classes to master's students of all four schools of the university in the ethics of technology, philosophy of science, sociology of technology and innovation, and critical thinking about science and technology. During the academic year 2021–2022, over 1500 students were enrolled in META courses (out of a total of

7900 master's students attending the Politecnico di Milano). In addition to these teaching activities, members of META also participate in (inter)national research projects, often with colleagues from their respective engineering, architecture, or design departments. META also carries out public outreach initiatives to foster critical thinking about science and technology among the general public in addition to organizing and hosting public lectures, workshops, debates, and occasionally, cultural events such as theater plays.

The personal narrative of my experience at the Politecnico di Milano encompasses two transitional moments, which each afford me a slightly different vantage point on the process of integration and the challenges involved. The first moment starts with my arrival in Milan in early 2021 to work as a postdoc affiliated with the META study unit while based in the Department of Management Engineering, which is a rather unique discipline that is particular to the historical emergence and institutionalization of engineering in Italy, but that can briefly be summarized as an interdisciplinary field that combines approaches from applied economics, behavioral economics, and management/organization studies. The second moment involves my transition from postdoctoral scholar to the position of assistant professor (junior-level, also known as "RTDa" in Italian bureaucratic parlance) with a three-year temporary contract. Although I am currently still based in the same department and (happily) affiliated with META, my promotion to a more senior position as well as the transition to a (slightly) more secure form of employment have provided me with insights and experiences that I could not have obtained as a postdoc, particularly on the level of conducting research, supervising students, and collaborating with engineering colleagues. Before turning to that aspect of being an integrated scholar, however, I will first discuss my experiences as an instructor.

3 Teaching as an Integrated Social Scientist

In the first year of my postdoc, I was asked to teach a course called the *Social Shaping of Technology* (hereafter *Social Shaping*), which I gladly accepted as an interesting opportunity to introduce engineering students to my home discipline of STS. Since my transition from postdoc to assistant professor, I am required to take on a specific teaching load, which means that I am now also teaching a second course, *Diversity-Aware Design of Technology Solutions* (hereafter *Diversity Aware*), which is co-organized with an engineering colleague from the Department of Electronics, Information and Bioengineering. These two courses are quite different in terms of their design, organization, and audience. *Social Shaping*, which was never offered at the Politecnico di Milano before my arrival, is "my" course, and I have full autonomy over its teaching materials, practical organization, student evaluation, and so on. *Diversity Aware* was also an entirely new course, namely a so-called "integrated" course that is co-taught by an engineer and a social scientist. Because of the considerable differences between these two courses and the challenges involved, I will discuss each one separately.

Social Shaping is, broadly speaking, intended to be a course that teaches engineering students about the social aspects of technology both in terms of its impact on society as well as its socially-constituted nature. The course has no prerequisites and open to master's students not just from Management Engineering, which is my home department, but also Electrical, Energy, Food, Space, Nuclear, and Aeronautical Engineering as well as Engineering Physics. In short, there is a wide range of engineering perspectives represented among the students, in addition to a broad diversity in terms of socio-cultural backgrounds and interests, which is further diversified by the presence of a significant number of international (and exchange) students. For many students, *Social Shaping* is often their first and perhaps only encounter with academic sociology, and therefore a prime opportunity to make them reflect on the practice of doing science and engineering and its social constitution and implications. However, students' unfamiliarity with sociological frameworks and concepts—and especially their inexperience with reading academic texts from the social sciences or humanities and their disciplinary jargon—means that lectures need to be, at least in part, dedicated to explaining the assigned readings rather than providing supplemental background materials and illustrations to course concepts.

In order to prepare students for reading sociological articles, I ask them to read Loïc Wacquant's *Practical Tips for Reading Sociology*, which provides several helpful guidelines. To ensure that the core concepts of the readings are grasped by students, I provide a comprehensive and non-jargon filled overview during my lectures, and supplement the academic readings with texts or videos that are produced for more popular outlets (e.g., Aeon.co, Logic Magazine, YouTube) or high-quality news media (e.g., New York Times, The Guardian, The Atlantic) that nonetheless provide critical discussions of recent developments in science and technology. Such shorter, often more provocative articles or videos can easily be read or watched by students in class, and then collectively discussed in a very productive manner. To foster class discussions, I make sure that students are given the time and space to formulate their thoughts—not always an easy feat for non-native English speakers—and try to provide clarification when necessary, which can be particularly tricky during complicated discussions in which arguments between students can occasionally become heated. Last, but certainly not least, I allocate a significant portion of the course to instructing students how to write an academic essay, which is the final assignment of the course. In fact, many students express interest in *Social Shaping* exactly because they know that its final assignment involves writing an essay, which they perceive as an important way of preparing for the master's thesis that they are expected to complete for obtaining their degree. Despite the necessity of writing a final thesis, the Politecnico di Milano does not appear to offer many courses that allow students to learn or improve their writing skills, which is why at least some of them turn to courses offered by META even if they are often quite fearful of needing to write a final essay.

The primary objective of *Social Shaping* is to disrupt dominant ideas about science and engineering as fields that have a one-way relationship with society in which scientific ideas and technologies have "effects" and "outcomes" but are otherwise unaffected by social, ethical, or political concerns or developments in society [9].

From the very beginning, there are two major and interconnected narratives that need to be challenged and reframed in order to work towards this objective. The first is to critique the linear model of knowledge production and technological development that draws a simplistic and unidirectional relationship between design, creation, diffusion, and usage [10, 11]. As an alternative, I try to teach students that technoscientific knowledge and objects are rarely, if ever, only developed in isolated labs without input or influence from non-experts and broader social beliefs and concerns. Instead, technoscience comes into being through complex two-way or even multiple ways of interaction between several social groups, which have different positionalities or interests and can exert varying degrees of power [12, 13]. That is to say, rather than merely thinking about technology as something that has impacts in society, I try to invert this conventional image and demonstrate how we can also think of society as being "inside technology" [14], or see science, technology, and society as "co-constructed" through mutual interactions [15, 16].

The second and tightly connected issue is to unsettle the often taken-for-granted relationship between dominant narratives of scientific development, technological solutionism, and social progress [17, 18]. Students commonly assume that scientific advancements will, in and of themselves, lead to social progress, and that technological developments are the primary solution to many social problems. A key initial step is to provide students with the tools to recognize such narratives and the types of progress they envision, and next, to make students ask who or what might be excluded from this conceptualization of progress. In short, who precisely is technological innovation for? Once the image of progress and the values and ideals it embodies, and for whom, have been deconstructed, students are much better able to recognize and critique the political power relations that motivate such images. However, a problem that frequently occurs while critically addressing progress narratives is that some students may interpret this endeavor as being anti science or anti innovation, or even worse, as betraying an extreme relativist position that endorses nihilism and hopelessness. Such students may raise tough questions such as, "Isn't it undeniable that technoscientific developments have raised the level of material wealth and comfort?", or they may point towards the fact that medical technologies have provided many with the opportunity to live healthier and longer lives (see also [1] on this specific issue). A key reply to such questions is not, of course, to deny that such forms of developments may count as a form of "progress", but rather to highlight that what is considered "development" or "progress" is something that evolves over time and that can differ greatly between cultural contexts, social groups, and individual people. In short, our notions of technoscientific progress are situated in a specific time and place, and therefore cannot be considered as single, timeless or universal [19–21].

On a practical pedagogical level, rather than treat technoscience as something that is done only within research and development institutions or technology companies and startups, I have opted to work with key concepts and case studies that provide insight into science and technology as systems of knowledge production and material development that are intimately connected not only to broader narratives of market imperatives, economic efficiency and productivity, and national identity, but also to ideals of democratic participation, the public good, and the expression of socio-cultural identity and values. Although the course focuses on technology rather than science, I often find it necessary to take brief detours into the history and philosophy of science to explain, for example, that all observation is theory-laden [22] or that seemingly "natural" differences in terms of sex, gender, and race are socially constructed and originating from very specific, situated historical developments in biology, medicine, and technology [23–27].

In terms of course materials, I move from a swift introduction to Sociology as an academic discipline (touching briefly on questions such as "what is society?" and the relationship between individual agency and social structures) towards broad and basic concepts that underpin the social construction of technology, such as "technological determinism", "technological solutionism", "socio-technical systems", and "opening the black box", which are helpful means of challenging or deflating common assumptions about the nature of scientific truths and the establishment of scientific consensus as well as the idea that technology might somehow be value free or impacting society in an independent and autonomous manner. After this, I introduce more focused and detailed approaches, concepts, and case studies. I usually start with introducing key frameworks of STS, such as the Social Construction of Technology and Actor-Network Theory by assigning (and lecturing on) classic texts by Trevor Pinch, Wiebe Bijker, Bruno Latour, and Langdon Winner. After this, I introduce students to more recent feminist and postcolonial approaches in STS (and more recent developments in technoscience)—by authors such as Donna Haraway, Judy Wajcman, Ruha Benjamin, and Virgina Eubanks—which focus on issues of gender, race/ethnicity, class, disability, and social inequality more broadly. This move from classical studies of technology and innovation (often based on relatively "old" technologies such as bicycles, cars, walls or doors, and bridges) towards newer technologies such as algorithms, AI models, and computer vision, is crucial for drawing a connection between early STS frameworks and contemporary technological developments, which students tend to find more directly relevant to their lives and about which they often want to write their final essays.

In its first year of being offered, *Social Shaping* attracted 119 students, which is a number that would have been considerably higher still if the course coordinator had not intervened and capped the course. In total, 110 students successfully completed this first edition of the course, and student feedback indicates that it was highly appreciated (and evaluated) as a welcome addition to the standard engineering curriculum. In short, there is a clearly an urgent wish among engineering students to be provided with a critical perspective on their research and work that differs from as well as supplements their usual courses. That being said, from the little experience I have with supervising master's students and their theses, it clearly remains a challenge for students who wish to pursue more sociologically-oriented research topics to study and write about such topics without falling back on the engineering frameworks, concepts, and methods that have dominated their curriculums so far. Even though the introduction and exposure to social science and humanities perspectives is generally appreciated by engineering students, it remains unclear if the impact of this exposure is more than merely superficial. That is to say, although META classes are ostensibly offered to students in order to steer their research and future work in a more socially-oriented direction, it remains ambiguous whether students are actually adequately prepared (and rewarded, either in terms of research evaluation as well as future job prospects) for pursuing such endeavors.

I will now turn to briefly discuss the second course in which I am involved as an integrated social scientist. The main difference between *Social Shaping* and the course that I co-teach with an engineering colleague—*Diversity Aware*—is that I am more obliged to demonstrate the utility of STS approaches in ways that fit into the shorter schedule and much more practical and applied orientation of the course. In other words, my contribution to the didactic contents of the course needs to be directly relevant to technological problems, or as the final word of the title of the course indicates—"solutions". My co-instructor is a full professor in "automatic control" in the Department of Electronics, Information and Bioengineering, and her main focus is on the design and evaluation of control technologies for automotive systems and smart mobility infrastructures. Despite this obvious technical orientation, her work seeks to directly engage with issues of social inequality, particularly in urban mobility, which is why she co-organized *Diversity Aware* to explore how mobility systems and services should consider diverse socio-economic variables and intersectional identities in relation to technology.

In a nutshell, the course is supposed to provide students with instruction in, first, the theoretical fundamentals of social identity and approaches to studying intersectional diversity, and second, the practical application of intersectional perspectives in the design and development of human-centric and diversity-aware smart mobility systems. The first half of the course is led by me and organized in a manner that is similar to that of *Social Shaping*, namely as a combination of lectures, class discussions, and in-class activities such as reading or writing assignments as well as roleplay exercises and debates. The second half of the course, which is led by my engineering colleague, is organized in a much more applied manner and focuses on the concrete design of inclusive mobility systems. Here, students are introduced to an engineering framework involving data-driven social-network analyses of adoption mechanisms as well as principles for feedback-based policy modelling and the development of human–machine interfaces based on nudge theory. In contrast to the sociological part of the course, which is evaluated by means of an exam, the engineering part of the course is evaluated by means of a group project and presentation in which students are expected to integrate and apply intersectional concerns and frameworks to concrete engineering projects.

One of the primary challenges for me as the social scientist of the course is to compress and deliver not only a brief introduction to the field of Sociology and social-constructivist approaches to technology, but also to provide an adequate introduction to the key course concept of intersectionality and the multiple types of identity, privilege, and inequality that this approach seeks to address. This forces me to abandon the more elaborate approach I take in *Social Shaping* and instead opt for a more applied way of approaching intersectionality. Rather than providing a broad historical and sociological perspective, I lecture on more concrete and practice-oriented approaches that have been developed by scholars in feminist, disability, and critical race studies, and particularly those that seek to address the relationship between design, power and social justice as well as redress structural inequalities. For example, the "Data Feminism" framework developed by Catherine D'Ignazio and Lauren F. Klein [28] provides a set of principles—and a plethora of vivid examples—that can guide students in the collection, analysis, and representation of various forms of data, particularly in the second part of the course. Similarly, Sasha Costanza-Chock's work on "Design Justice" [29] is extremely helpful for problematizing universalist design principles and demonstrating how the inclusion of marginalized communities can be a crucial means of making design practices more equitable.

It is crucial that the sociological and qualitative part of the course is delivered in the first half of the course rather than the second. This sets the tone of the course in an important manner in the sense that the messiness of social reality—and how this can be understood from a qualitative rather than a quantitative perspective—is confronted from the very beginning rather than introduced as an afterthought as often tends to be the case when addressing the social implications of technologies and infrastructures. That being said, it remains to be seen—as *Diversity Aware* has been taught only for the first time at the time of this writing—how much of the ideals of intersectionality and social justice can be upheld and applied when confronted with the hard reality of quantitative approaches, practical design imperatives, and cost–benefit analyses. Nevertheless, the challenge of finding a consensus in terms of pedagogical approaches to teaching about intersectional design is something that both co-instructors of this course were prepared to confront from the onset of the course. Moreover, this course is clearly already the start of a fruitful process that will teach us a lot about ourselves and our respective disciplines. It therefore embodies both the tensions and pedagogical ideals that the integration of engineering and social science perspectives in a single course are bound to produce.

4 Researching and Collaborating as an Integrated Social Scientist

I arrived at the Politecnico di Milano as a postdoc in early 2021 to work on a project that examines the role of public participation in scientific research through digital platforms. My affiliation would be with the META study unit, but I would be based in the Department of Management Engineering and expected to report to and collaborate with my colleagues there. In the middle of 2022, my postdoc was fortuitously cut short by a promotion to the position of assistant professor in the same department, which expanded my abilities to supervise students and collaborate with engineering colleagues. These two periods have provided me with insights into the practical and interpersonal challenges that emerge from interdisciplinary collaboration as well as the organizational and institutional challenges that an integrated social scientist may face while working in an engineering university.

My first experience with interdisciplinary collaboration took place when setting up my postdoc research project. The professor I was going to work with is an applied economist who had conducted previous research on citizen science platforms, entrepreneurship in innovation, and science funding mechanisms. Although I was quite uncertain and even mildly worried about the prospect of working with a quantitative researcher, I was also curious to embark on a totally new endeavor and excited about the prospect of learning quantitative and computational methods. I imagined, perhaps naively, that I would be able to relatively rapidly expand my skillset as a researcher and fruitfully collaborate with my quantitatively trained colleagues.

After a series of conversations with my postdoc supervisor, we decided that I would conduct a qualitative study of a crowdfunding platform for scientific research [30], while jointly working on a mixed-method study that focused on the presence of campaigns that seek support for "undone" or "grassroots" research as well as alternative science or pseudoscientific projects in a dataset that would be scraped from the same crowdfunding platform. After collecting this data (with the assistance of someone with the computational skills that I sorely lacked), I was confronted with the harsh reality of an Excel spreadsheet with thousands of data points that, by necessity, needed to be analyzed in a quantitative manner. My collaborator initially suggested using a relatively straightforward word-analysis software called Linguistic Inquiry Word Count (LIWC)—frequently used in psychometrics as well as in computational studies of sentiment [31]—which assigns to a particular word a range of emotive dimensions and scores them according to their usage. However, LIWC is based on a number of assumptions about the use of language and its meaning that run counter to all my instincts as a qualitative researcher. One issue is that words are always indexical and can be polysemous, which means they can take on different meanings depending on the context in which they are written as well as the other words that surround them in a text. Moreover, the meaning of terms is always contingent on their interpretation by human subjects, which further underscores that the meaning of terms can be as variable and complex as humans are [32]. Another issue concerns the categorization of words and the assignation of psychological traits. It is one thing to assume that neuroticism is a human psychological trait, but another to argue that using the first-person pronoun is a way of signaling this, and yet another altogether to connect this to the likelihood of upholding pseudoscientific beliefs or having revolutionary intents. In other words, such assumptions set my alarm bells ringing as an ethnographer for whom context is everything, not just something that needs to be reduced even in exchange for the benefits of a large-scale aggregated analysis. "Surely the ambiguity of social life cannot be flattened into binary categories and measurable types that say just as much about our assumptions as researchers than our research subjects?!", or so I thought.

My encounter with the challenges of conducting mixed-methods research and interdisciplinary collaboration is, of course, not just a personal issue between researchers with different backgrounds. Although our disciplinary differences and the discussions that these gave rise to highlight how the process of integration involves practical as well as interpersonal challenges that play out in face-to-face interaction with colleagues, it is also indicative of a much broader issue, namely the difficulty of crossing, let alone breaking down, the barrier between qualitative and quantitative ontologies and epistemologies in the study of science and technology [33]. The challenges posed by mixed-methods research draw out precisely the difficulty of interdisciplinary collaboration between fields that have fundamentally different theoretical and methodological assumptions. To be clear, this is not to suggest that qualitative approaches are somehow more insightful or "better" than quantitative approaches, but instead to highlight how tricky it is to collectively determine research questions as well as techniques for answering them that are mutually comprehensible and acceptable to all parties involved. For example, my initial disconcertment [4] and objections to the proposed psychometric approach have drawn out the collaborative process to some extent, and indeed this is still ongoing. Slowing down the research process in this manner can be seen as a drawback, particularly considering the current pressures to publish rapidly and repeatedly, but it can also be seen as a valuable opportunity to think through the advantages and disadvantages of both qualitative as well as quantitative approaches, the opportunities, limits and drawbacks that their methodological frameworks offer, and the necessary tradeoffs involved in terms of the types of insights that each approach can provide. Although this is a challenging process that requires patience, generosity, and flexibility from both parties, I continue to believe that our analysis of this particular dataset will be all the better for having gone through this process. Moreover, it will also allow us to reconsider our taken-for-granted disciplinary assumptions and to refine our research methods in ways that will retain the sensitivity to context and subjectivity that qualitative research demands while simultaneously benefitting from the scope and scale that quantitative research affords.

Thankfully my relationship with this particular colleague as well as others in my department remains strong as we continue to discuss and collaborate, but disciplinary differences can also lead to more awkward and uncomfortable moments. As a recent hire in my department, I was encouraged to attend the main conference of management engineers in Italy in order to get a sense of the disciplinary culture and to familiarize myself with the major topics and methodologies of this field. With some reservation but also some curiosity, I submitted an abstract to the conference that was based on my qualitative study of the aforementioned crowdfunding platform, and was accepted to present at the conference and submit a full paper that would be given feedback on by a discussant. After delivering my presentation, which, to my mind, gave a nuanced and carefully considered critique of crowdfunding for scientific research and its potential for exacerbating inequalities among scientists, I was met with a series of comments that, although delivered in a cordial way, questioned the basic premises of the qualitative approach and its ability to provide relevant or interesting insights. "Why didn't you try to analyze these dynamics in a more measurable and objective manner?". Although this critique of my approach was perhaps not entirely unexpected, it was nevertheless dispiriting and served to highlight my position as an outsider who is not aware of the "rules of the game" [34].

Such practical and interpersonal challenges, which require a considerable degree of negotiation but are generally manageable, can be insightful and ultimately fruitful.

However, the integrated social scientist can also be faced with organization and institutional challenges that are often much harder to negotiate and circumvent. My position as a META scholar at the Politecnico di Milano is explicitly supposed to fulfil the function of "outsider" social scientist who is supposed to make the engineers aware about the social dynamics and implications that are inherent to their work. However, my actual appointment is in the Department of Management Engineering, and like my other META colleagues who are embedded in other engineering departments, we are also expected to contribute to departmental goals and duties that our engineering colleagues are obliged to do, such as supervising students and improving our department's profile both within the university as well as beyond. In other words, in theory I am expected to maintain my disciplinary identity as a social scientist, but in practice I am situated within an engineering department and expected to contribute to its institutional and disciplinary profile by attending conferences and publishing in journals that contribute to my departmental output and ranking. Being embedded in an engineering department in such a manner is an important means of "staying close to the action" [35] as well as intervening through teaching and collaboration, which is a rare opportunity for a social scientist of science and technology. However, it also gives rise to important drawbacks and challenges, particularly in terms of the types of research that can be done and how these are evaluated and rewarded.

First, my position as an insider/outsider forces me to strategize and publish in journals that are relevant to both my department and my own disciplinary identity and field. Fortunately, the disciplinary field to which my department belongs and the list of core journals in which members of this field are expected to publish contains some overlaps with the disciplinary field of Sociology to which I belong according to the Italian Ministry of Education. This is mostly due to the fact that many management engineers are, as applied economists and scholars of management and organization, interested in publishing in some social science journals, but this is certainly an exceptional situation that would have been far less likely to occur if I were embedded in, say, the Department of Nuclear or Mechanical Engineering. Moreover, my department allocates research funds on the basis of bibliometric indicators that not only privilege journals that are not my core disciplinary priority (if relevant to me at all) but also rely on a level of publication productivity that is extremely difficult, if not impossible, to achieve for someone whose research is fundamentally slower in nature. This dual and somewhat conflicting obligation to contribute—to the best of my ability—to two different disciplines simultaneously means that I have to prioritize trying to publish in certain journals that I might not be particularly appropriate for given my particular background and training. Moreover, it also means that publishing in some journals might mean that I am only fulfilling one part of my double obligation, which can be a tricky tradeoff to make given the time and resources necessary to publish even a single research article even if it is in a "top" journal in the field of either Sociology or Management Engineering. In other words, the institutional setup for this mode of integration in engineering departments imposes considerable challenges and constraints on the social scientists whose priorities are now split between two academic disciplines as well as two institutional goals, namely of providing social science expertise as an "outsider" (and maintaining my status as a scholar in the field of Sociology) while also functioning on a departmental level as an "insider" who contributes to the productivity and visibility of the department.

The issue I have described above is, of course, not unfamiliar to those who engage in interdisciplinary research and collaboration or who are embedded in departments or institutions that do not reflect their own disciplinary training. The necessity of bridging the divide between theoretical languages, methodological approaches, and epistemic goals makes it difficult to identify common research problems and is thus a major barrier to integration [5]. The integrated social scientist or humanist, especially when they are the sole outsider in their department or group, can easily find themself in a somewhat isolated and asymmetrical relationship with their colleagues in which their specific expertise is only a drop in the ocean of the hegemonic consensus, and furthermore, a drop that is added only after the main research questions and approaches have already been decided upon [2]. Overcoming such barriers to interdisciplinary collaboration requires significant investment in time and resources as well as organization commitment and careful intention to the different incentive structures and publication strategies pertaining to the identity of integrated scholars. Above all, as others have argued [5], integrated researchers should not only be brought on board to address one particular problem or to fulfil a certain mission (e.g., "social impact" or "human-centric technology"). Instead of such an instrumental notion of integration, it is perhaps better to develop 'integrative environments' that seek to foster a more complex and reflexive understanding of real-world problems and that involve a much stronger institutional commitment to long-term interdisciplinary collaboration. Such an endeavor, of course, also requires addressing the much broader and more difficult issue of alignment between research institutions, funding agencies, government policies, and academic communities of multiple disciplines.

A second organizational and institutional challenge to integration concerns the precise nature, quality, and timing in which the input and expertise of social scientists is sought. A recent example is a large, five-year, multi-million project that my department has started as the result of being allocated a seal of "excellence" from the Italian Ministry of Education. This project—called *HumanTech*—focuses on research into the relationship between technologies and people in the transition towards more sustainable, inclusive, and harmonious digital and industrial systems. Despite widespread references to terms such as "human-centered" and "people-friendly" technologies as well as "collective wellbeing", this initiative was set up without any input from myself nor other members of META, who are ostensibly supposed to be experts in precisely this—the relationship between humans and technologies or how social processes and concerns as well as technoscientific research and development are co-constructed. This means that important decisions about how this project will be organized, the research infrastructures that will be set up, the new faculty members hired, the evaluation of research outputs and the goals that will be achieved have been made without a say by those who are ostensibly hired by the university to ensure that such projects are indeed tailored towards social and ethical concerns. This begs a number of important questions: when precisely should integration take place? What criteria are there for evaluating the degree and quality of integrative collaboration and its outcomes? What aspects and actors should be subject to integration? What precisely are the issues that require the changes that the integration of perspectives from the social sciences and humanities are supposed to achieve? How might one evaluate the failure to integrate? And what institutional or disciplinary incentives and rewards are there for integration and collaboration? Without explicit discussion and acknowledgement of these questions, many of the underlying assumptions that drive the turn towards integration at the Politecnico di Milano remain relatively implicit and invisible. Indeed, as long as these questions remain unaddressed and unanswered, and as long as the expertise in the social sciences and humanities provided by the META study unit is not integrated in a more formal manner, it is tempting to say that this form of "integration" is more a gesture towards a laudable but impractical ideal rather than a substantive change from previous practices.

5 Conclusion

This chapter has outlined some of the opportunities and challenges involved in enacting "integration" as a social scientist at a technical university. I first focused on the process of teaching engineering students, and some of the challenges encountered and pedagogical techniques used to introduce students to critical thinking skills as well as a more socially-informed perspective on technology. I cannot deny that this effort is, at times, as much a challenging process as it is a delightful one. There is hardly a class that goes by without me being forced to reflect not just on the world of engineering but also my own disciplinary assumptions and teaching methods.

In terms of teaching, the "integration" of the social sciences in engineering seems to be a success. META classes are well-attended, popular, and generally appreciated by the engineering students, as can be witnessed from the positive (anonymous) student reviews as well as individual feedback in private. Even though there is still much room for improvement, it is clear to me that most engineers leave my courses with a more complex understanding of technoscience and its socially-constructed character. Although a portion of the final papers remains disappointing in terms of their critical depth—for example, some students continue to think that a psychological approach to technology is the equivalent of a sociological one—many students nonetheless leave the course with a more profound understanding of co-constructed nature of technology and society as well as the role they play in this process as engineers.

In terms of collaboration with engineering colleagues, the process of "integration" is more difficult, and the outcomes are, for now at least, less optimistic. Crossing or breaking down disciplinary divides between quantitative and qualitative approaches remains a daunting challenge. On the one hand, much work remains to be done on the part of myself and my colleagues at META in terms of translating the critical perspectives on technoscience from the social sciences and humanities and making these insights more accessible and perhaps even operationalizable for our engineering colleagues. Indeed, the field of STS as well as research/technical universities more broadly should consider their role in this task as well, given both the complexity as well as the urgency of addressing grand and intractable challenges such as, for example, climate change. On the other hand, it is not unfair to expect that engineers also seek to reflect on and rethink their own disciplinary assumptions and approaches in addition to attempting to integrate social, ethical, and political concerns into their engineering practices. The course that I am co-teaching with an engineer, for example, is a promising instance in which the integration of social concerns into engineering pedagogy and practices is starting to be done, but this can and should be done at a much larger and formal scale than is currently being done.

Challenges that hamper integration are not just found on the level of interpersonal and practical interaction but also on the level of organizational and institutional arrangements. The individual integrated social scientist is, to some extent, under pressure to fulfil a dual role that is difficult to achieve to a satisfactory degree. This dual role not only imposes a higher burden of labor but also runs the risk of pushing the social scientist into a more marginalized position both within their home discipline as well as the engineering department in which they are embedded. The issue of isolation and marginalization does not merely apply to the individual social scientist or humanist, who not only risks remaining at the periphery of their department as well as their own academic discipline, but indeed to the qualitative social sciences and humanities more broadly. The expertise that these perspectives have to offer and the types of knowledge they produce are often not amenable—because of their time-consuming, uncertain, and often exploratory as well as deconstructive approaches—to current research and funding programs that focus on pre-defined and carefully circumscribed "problems" or "missions" whose definition and interpretation tends to be dominated by the natural sciences of quantitative social sciences. Furthermore, it is also clear that the process of integration is relatively underdetermined and therefore somewhat lacking in terms of formalized institutionalized commitments that would allow for a more effective implementation as well as evaluation of the integrative ideal.

The perspective that I have provided here is, however, a partial one, and does not necessarily reflect the experiences of my META colleagues who are embedded in other engineering or design departments. Nevertheless, I can share that at least some of the doubts and frustrations that I have articulated here are shared by several of my younger colleagues who are in similarly precarious positions at an early stage of their careers while trying to fulfil what at times appears as a double duty. That being said, the acknowledgement that integration can be difficult and frustrating also has to be accompanied with the observation that direct interaction with engineers and the possibility of intervening in their practices is an invaluable opportunity not just to further make the case for integration but also to further reflect on our own positionality, methods, and disciplinary assumptions. Reflexivity is a core tenet of STS after all.

Moreover, there are a number of important issues that I have not had the opportunity to discuss, such as the considerable challenge of supervising students who are writing theses using methods and techniques that I am not in a position to adequately evaluate, or the difficulties involved with attracting PhD students from the humanities and social sciences who do not wish to complete their dissertation under the aegis of an engineering department. Furthermore, I have not provided suggestions for potential solutions for improving the process of integration, which would probably require another chapter. That said, three very brief remarks on this can be made. The first is that the institutional and organizational identity of integrated social scientists and their role needs to be formalized in order to safeguard their primary function as the provider of a critical "outsider" expertise that does not have the potential to affect their research or employment. Next, it would be prudent to formalize the process of integration and its expectations, processes, and outcomes so that both social scientists and engineers can be held to shared standards for evaluating the success or failure of the integrative process. Finally, there is an invaluable opportunity at the Politecnico di Milano that is currently underexploited, which is to embed social science and humanities research at the level of laboratory practices. This would allow members of META not only to conduct fieldwork within the labs of the Politecnico di Milano—thus directly studying the engineering work that takes place therein and contributing to the well-established field of "laboratory studies" in STS—but also provide the engineering departments at the Politecnico di Milano with an opportunity to be provided with critical insights that are derived directly from their own local practices, habits, and beliefs. Such a form of integration—in the form of face-to-face interaction between social scientists, engineers, and their respective techniques and technologies in the messy reality of laboratory life—would truly live up to the ideal of integration and undoubtedly be beneficial to all actors, disciplines, and institutions involved.

References

1. Downey GL (2008) The engineering cultures syllabus as formation narrative: critical participation in engineering education through problem definition. Univ St Thomas Law J 5(2):428–456
2. Viseu A (2018) Integrating is caring? Or, caring for nanotechnology? Being an integrated social scientist. In: Subrahmanian E, Odumosu T, Tsao J (eds) Engineering a better future. Springer, Cham, pp 141–166
3. York E (2018) Doing STS in STEM spaces: experiments in critical participation. Eng Stud 10(1):66–84
4. Smolka M, Fisher E, Hausstein A (2021) From affect to action: choices in attending to disconcertment in interdisciplinary collaborations. Sci Technol Human Values 46(5):1076–1103
5. Pedersen D (2016) Integrating social sciences and humanities in interdisciplinary research. Palgrave Commun 2(16036):1–7
6. Daniel KL, McConnell M, Schuchardt A, Peffer ME (2022) Challenges facing interdisciplinary researchers: findings from a professional development workshop. PLoS ONE 17(4):e0267234
7. Lynch M (2000) Against reflexivity as an academic virtue and source of privileged knowledge. Theory Cult Soc 17(3):26–54
8. Spindler M, Booz S, Gieseler H, Runschke S, Wydra S, Zinsmaier J (2020) How to achieve integration? In: Gransche B, Manzeschke A (eds) Das geteilte ganze. Springer VS, Wiesbaden, pp 213–240

9. Williams R, Edge D (1996) The social shaping of technology. Res Policy 25(6):865–899
10. Bush V (1945) Science—the endless frontier. A report to the President. United States Government Printing Office, Washington, D.C.
11. Rogers EM (1962) Diffusion of innovations. Free Press of Glencoe, New York, NY
12. Funtowicz S, Ravetz J (1993) Science for the post-normal age. Futures 25:739–755
13. Gibbons M, Limoges C, Nowotny H, Schwartzman S, Scott P, Trow M (1994) The new production of knowledge: the dynamics of science and research in contemporary societies. Sage, Thousand Oaks, CA
14. Winner L (1980) Do artifacts have politics? Daedalus 109(1):121–136
15. Cowan RS (1983) More work for mother: the ironies of household technology from the open hearth to the microwave. Basic Books, New York, NY
16. Jasanoff S (2004) States of knowledge: the co-production of science and the social order. Routledge, New York, NY
17. Marx L (1987) Does improved technology mean progress? Technol Rev 32–41
18. Johnston SF (2018) Alvin Weinberg and the promotion of the technological fix. Technol Cult 59(3):620–651
19. Edgerton D (2006) The shock of the old: technology and global history since 1900. Oxford University Press, Oxford
20. Haraway D (1988) Situated knowledges: the science question in feminism and the privilege of partial perspective. Fem Stud 14(3):575–599
21. Marx L (2010) Technology: the emergence of a hazardous concept. Technol Cult 51(3):561–577
22. Hanson NR (1958) Patterns of discovery. Cambridge University Press, Cambridge
23. Adas M (1989) Machines as the measure of men: science, technology, and ideologies of western dominance. Cornell University Press, Ithaca, NY
24. Benjamin R (2019) Race after technology: abolitionist tools for the new Jim code. Polity, New York, NY
25. Epstein S (2007) Inclusion: the politics of difference in medical research. Chicago University Press, Chicago, IL
26. Martin EL (1991) The egg and the sperm: how science has constructed a romance based on stereotypical male-female roles. Signs: J Women Cult Soc 16:485–501
27. Schiebinger L (1986) Skeletons in the closet: the first illustrations of the female skeleton in eighteenth-century anatomy. Representations 14:42–82
28. D'Ignazio C, Klein LF (2020) Data feminism. MIT Press, Cambridge, MA
29. Costanza-Chock S (2020) Design justice: community-led practices to build the worlds we need. MIT Press, Cambridge, MA
30. Hesselbein C (2023) Kickstarting science? Crowdfunded research, public engagement, and the participatory condition. Sci Cult. https://doi.org/10.1080/09505431.2023.2222741
31. Stark L (2018) Algorithmic psychometrics and the scalable subject. Soc Stud Sci 48(2):204–231
32. Sperber D, Wilson D (1995) Relevance: communication and cognition. Wiley-Blackwell, Cambridge
33. Kang D, Evans J (2020) Against method: exploding the boundary between qualitative and quantitative studies of science. Quant Sci Stud 1(3):930–944
34. Bourdieu P (1984) Distinction: a social critique of the judgement of taste. Harvard University Press, Cambridge, MA
35. Goffman E (1969) Where the action is: three essays. Allen Lane, New York

Future Medicine: Towards a More Conscious and Ethical Communication

Beatrice Lucia Bona

1 Introduction

Over the past three decades, the application of nanotechnology into medicine, known as nanomedicine, has emerged as an extremely promising field. Nanomedicine holds great promise for transforming disease screening, diagnosis, and treatment at both the population and individual levels, thereby greatly impacting human health [1, 2].

Nanotechnology harnesses the unique properties of nanomaterials, which exhibit different characteristics compared to their larger counterparts. These materials, typically ranging in size from 1 to 100 nm, have the ability to interact and integrate with biological systems. They can travel through the bloodstream, penetrate tissues and cell membranes, enabling the development of nanoparticles that function as "shuttles" for targeted drug delivery [1, 2]. By engineering nanoparticles, it is possible to enhance drug specificity, pharmacokinetics and efficacy while reducing the overall toxicity and side effects compared to traditional drug administration methods. If properly designed, nanoparticles can act as "Trojan horses," protecting therapeutic agents from degradation by enzymes present in the human body and delivering the drug at the required dose precisely to the intended pathological site.

Research in nanomedicine is currently underway worldwide, focusing on the development of nanotechnology applications for preventing, diagnosing, and treating various diseases such as heart conditions, neurological disorders and cancer, as well as for tissue regeneration approaches both in vitro and in vivo [1].

Since the first approval of a nanotherapeutic for clinical use[1] by the Food and Drug Administration (FDA) in 1995, nanomedicine has experienced a significant

[1] Doxil®, a nanodrug based on a lipid nano-vector containing a chemotherapeutic drug.

B. L. Bona (✉)
Department of Chemistry Materials and Chemical Engineering "Giulio Natta", Politecnico di Milano, Milan, Italy
e-mail: beatricelucia.bona@polimi.it

S. Chiodo et al. (eds.), *Improving Technology Through Ethics*,
PoliMI SpringerBriefs, https://doi.org/10.1007/978-3-031-52962-7_3

boost. Today, advancements in nanodrugs, nanoimaging, and theranostic systems[2] are expected to revolutionize the future of medical practice [3].

However, it is only recently that nanomedicine has garnered widespread attention. In late 2020 and early 2021, regulatory agencies such as the FDA and the European Medicine Agency (EMA) granted authorization for the large-scale commercialization of messenger RNA (mRNA) based nano-vaccines to combat the COVID-19 pandemic [4]. This event marked the transition of mRNA technology and nanomedicine [5] from a concept mostly restricted to the research and clinical communities to a public domain. While this breakthrough held tremendous potential for medical innovation, it also raised significant ethical and safety concerns [3, 6].

It is not new to encounter ethical and social issues when scientific advances, especially those affecting public health, emerge [6, 7]. Nanomedicine is at the beginning of its journey towards widespread adoption and given the limited understanding of nanomaterials' risks to humans, it is crucial to develop regulations to control their use and address their ethical and social implications [8]. Responsible engagement with humanity and anticipation of the social consequences of technological innovation are vital to prevent harm, misuse, and misinformation [9].

When introducing something "new" in the medical field, understanding and addressing its potential social impact is complex. It involves not only assessing the direct risk factors associated with a new therapeutic or treatment methodology but also evaluating potential damage in a broader sense. Disinformation, lack of transparency, and inconsistent communication can have a profound impact on public health, mental well-being, and trust in medical innovation. The recent COVID-19 pandemic serves as a prime example, where confusing communication from experts and governments amplified social fear and gave rise to a multitude of "self-proclaimed experts" disseminating misleading information and unfounded opinions, ultimately leading to negative social consequences, and jeopardizing public health [10, 11].

This highlights the critical importance of ethical considerations throughout the research process and the introduction of innovations to society [12]. Prioritizing the development of the next generation of ethically aware experts and addressing existing ethical gaps among researchers and doctors is essential. Simultaneously, experts should strive to inform and raise awareness among the public about medical innovation through transparent information. Patients, their families, and society as a whole should be fully informed about the benefits and risks of nanomedicine, enabling them to understand the potential implications it may have on their health and make informed decisions. To achieve this, researchers, governments, and pharmaceutical companies should promote and support the development of campaigns and programs that facilitate transparent, simple and open discussions about the ethical and social impact of nanomedicine.

The history of innovation has taught us that a lack of information often leads people to perceive new technologies as dangerous or disruptive. Therefore, experts should

---

[2] Theranostics is a term formed by the word "therapeutics" and "diagnosis". Theranostics systems are nanoscale platforms able to integrate both diagnosis and treatment of a specific pathology.

prioritize clear communication to foster trust in new technologies and protect individuals from misleading and baseless information [6, 7]. An effective medical progress can be pursued only by achieving a high level of ethical awareness and cultivating a clear, critical, and transparent communication within the scientific community and between experts and the general public [12, 13].

In the following sections, the role of ethics and communication in medical innovation will be analyzed, aiming to raise awareness among researchers about the importance of recognizing their ethical and social responsibilities and the influential power their opinions hold in society. This is crucial for shaping fair and transparent progress while alleviating public fears, prejudices, and reactions towards innovation.

2 Analysis of a Case Study: The Hidden Danger Behind Non-transparent Forms of Communication

In August 2022, while on vacation in North America, I found myself reflecting on the upcoming course I would be joining about the ethical and social impact of emerging technologies. I was both excited and apprehensive as it meant stepping out of my comfort zone. With a limited understanding of ethics in science, I felt the need to enhance my knowledge on this crucial topic, particularly in the context of medical research. I wanted to take a small initial step towards becoming a more responsible and well-rounded researcher, both scientifically and at a human level.

As I was lost in thought, I noticed a billboard advertising enrollment in a Phase I clinical trial for testing a new treatment for neurological disorders in humans. Initially, I was impressed and saw it as a way to raise awareness about the importance of clinical trials for medical advancements. It seemed progressive to promote clinical trials so openly. However, upon closer inspection of the billboard, my initial impression changed. The trial's purpose, the specific pathology and treatment being addressed, and the significance of participating in terms of contributing to science were given very little attention (only a couple of lines in small font in the background). Instead, a large picture of smiling people hugging each other dominated the billboard, accompanied by a prominently displayed promise of a competitive reward for trial participants. The level of biased information in that advertisement was concerning to me, highlighting a lack of ethical and transparent communication.

Leaving aside the broad and complex debate surrounding the ethics of using incentives for clinical trial participation [14], what is unequivocally unethical in this context is the prioritization of payment information over other trial details through misleading formatting [15]. In this advertisement, payment information is given undue prominence, exploiting people's vulnerabilities, and relying on economic incentives as the primary means to capture their attention and influence their decision. Furthermore, visually associating the clinical trial process with a picture of smiling people conveyed a confusing message. Clinical trials represent hope for patients with serious or incurable conditions and their families. While I agree that people should be willing to participate and contribute to science, it is important to acknowledge that clinical trials can be complex, potentially painful, and not entirely risk-free. Therefore, depicting people smiling as if they had just undergone a simple cosmetic procedure seemed inappropriate. Subtle advertising tactics can influence people's decision-making, potentially leading them to make choices that are not in their long-term best interest. People should choose to engage in a clinical trial only after fully understanding the process, its implications in terms of risks and benefits, both for themselves and society, and not due to manipulative advertising.

Before continuing further, let me briefly explain what clinical trials are and how they are regulated. The clinical and commercial approval process for a new therapeutic agent typically involves multiple steps, lasting around 10 years, unless there is an emergency use authorization. Clinical trials represent the final stage of this process. A typical research program focuses first on developing a new drug with the desired properties, followed by evaluating its effects on biological systems in terms of therapeutic efficacy, toxicity, and safety. This evaluation occurs in vitro on cells, then in vivo on animal models, and finally on humans. Clinical trials assess whether a new therapeutic approach is superior to existing ones or to a placebo (if no alternative treatments are available). Their ultimate goal is to ensure the safest and most effective treatments for specific pathologies.

The clinical trial process articulates into four subsequent phases, each one addressing a different question [16, 17]:

- Phase I: is the new drug safe?
  The drug is tested on a small number of healthy people (except for oncological studies in which patients are involved) by gradually increasing the dose to identify potential side effects.
- Phase II: does the new drug work and at which dose?
  The drug is tested on a larger group of ill people to define the minimal effective dose.
- Phase III: is the new drug the best available one (better than the one used until that moment)? The drug is tested on a larger group of ill people and compared against an existing treatment or a placebo.
- Phase IV, pharmacovigilance: what happens after the new drug is commercialized?
  The safety, side effects and effectiveness of the medicine continue to be studied also after commercialization.

Clinical trials are extensively regulated internationally. The protocol must be approved by the relevant authorities in each country and by Independent Ethics Committees comprised of scientists, healthcare professionals, and citizens. Additionally, clinical trials must comply with Good Clinical Practice (GCP), an international standard that governs scientific ethics and quality, outlining the planning, execution, recording, and reporting procedures of human clinical experiments. Compliance with this standard ensures data reliability and the protection of participants' rights, safety, health, and well-being [16]. To enroll in a clinical trial, individuals undergo a detailed interview and screening conducted by a board of experts. They also read and sign an informed consent form, guided by clinicians and researchers who explain the trial's potential implications, risks, and benefits. A list of questions is typically provided to address any doubts and concerns that participants may have [16].

Considering all these aspects, one might wonder why misleading advertisements like the one mentioned should be considered an ethical issue with potential effects on people's health. Let's consider part of the definition of informed consent according to the Clinical Trials Directive: "a decision, which must be written, dated, and signed, to take part in a clinical trial, taken freely after being duly informed of its nature, significance, implications, and risks..." [18]. The objective of the informed consent is to empower potential participants to make voluntary, uncoerced decisions about whether to participate in a trial or not [19]. However, if the consent process actually begins with the recruitment of people through the use of misleading advertisements can we truly consider the final decision to be entirely voluntary and uninfluenced? How informed can the consent be in a completely transparent manner if initial recruitment involves advertisements that subtly influence people's minds, playing on their weaknesses while concealing the actual risks and implications of clinical trials behind images of smiling people?

Experts in the field who come across such advertisements know precisely what the Phase I clinical trial process entails. They know that scientists do not proceed to the trial stage without evidence of the new drug's effectiveness [20]. However, they also understand that the real risks and effects on humans of a new treatment are not entirely known or predictable at the start of a clinical trial. Experts are aware that experiencing potentially debilitating side effects during the trial is highly possible. Nonetheless, they also recognize that participating in trials contributes to science and medical innovation, offering hope to patients and their families, paving the way for a potentially better future for generations to come [21]. Experts possess the knowledge and tools to comprehend and evaluate the information in advertisements, even if it is partial and misleading. They may consider the promise of an economic reward as one factor in their decision, but they do so consciously and with balance. On the other hand, common individuals without expertise in the field, especially those who are vulnerable or facing economic difficulties, are defenseless against misleading communications. They become easy targets for influence. The economic reward may become their sole motivation, potentially overshadowing the need for a comprehensive understanding of the trial's risks and implications for their health. Consequently, ordinary individuals may provide inaccurate information in order to remain enrolled in the trial, acting against their long-term interests. This can lead to delays in the evaluation process and compromise the reliability of data.

In such a scenario, the ethical foundations of clinical trials, as well as the general ethical principles of respecting human beings, are undermined by non-transparent communication that promotes a game of mutual interest. On one side, to recruit enough participants, the trial-conducting entity influences citizens' decision-making processes through biased advertising. On the other side, citizens are subsequently induced to perceive the trial solely as a source of income. This creates a dangerous dynamic that, although sporadic and isolated, has the potential to significantly compromise the scientific and social significance of clinical trials. Years of efforts to raise awareness about the importance of research and medical innovation could be undermined, eroding social credibility, perception, and trust in the long run.

Without a proper ethical culture among all players in the scientific community, the dangers posed by biased communication, such as the one described, on individuals and their perception of medical innovations in the short and long term can be underestimated. If we were to think solely from a non-ethical standpoint, merely aiming to achieve the scientific objective of testing a new therapy, people could be surreptitiously induced to undergo an experimental treatment during the clinical trial stage using misleading communication. But what happens afterward? What about when the new treatment is commercialized, especially if it is something extremely innovative? How can scientists convince society that the treatment is safe if they haven't established a transparent communication channel from the beginning, if they haven't adequately sensitized people about the process timeline, controls, scientists' dedication, and the benefits for society?

History teaches us that major innovations are frequently met with social rejection driven by fear of the unknown, uncertainties, and risks associated with change, especially in the context of medical advancements that have a direct impact on people's life [22, 23]. To overcome this, constant and widespread use of transparent, coherent, simple, and accurate communication, along with outreach policies, is essential for the scientific community to establish trust with the population. This fosters informed acceptance of medical innovation, reduces fear and resistance, and ensures that the progress of future medicine aligns with fairness and ethical principles [12, 13].

Scientists and experts should be acutely aware of their ethical [24, 25] and social roles [26]. Clear, transparent, and effective communication is vital for mitigating panic, encouraging appropriate behavior, and framing the progress of future medicine at each stage in a manner that upholds fairness and ethical principles [27].

3 Evolution and not Revolution for a Stable Ethical Progress

One day, during a discussion about the role of ethics in new technologies, a professor who serves as an example and inspiration to me shared a concept that stayed in my mind and changed my perspective on ethical dilemmas. She said, "stable ethical progress in any field cannot be achieved through revolution but through a gradual evolution process." This highlighted the often-underestimated importance of shaping progress through small actions that have long-term effects, allowing ethical concepts to be integrated slowly into our being over time. To achieve a higher level of ethical consciousness, we need a broader approach that involves constant and long-lasting sensitization and education at different levels, from the complex scientific environment of experts to simpler communication for the general public.

For instance, ethics concepts should be better integrated into different courses throughout the academic journey, including bachelor's, master's, and PhD programs, to cultivate a next generation of researchers and doctors who are more ethically aware [28]. Currently, in most Italian universities, students have the option to include ethics courses in their curriculum, leaving the decision up to them. However, due to the existing lack of general awareness about these topics, students often prioritize more technical courses that they perceive as more useful. That is why, until we reach a higher level of general ethical consciousness in society, courses on ethics and its implications in science should be made mandatory in universities. Additionally, open discussion sessions involving students and various stakeholders, such as experienced researchers, company managers, and professors, should be encouraged to promote the importance and implications of ethics in the innovation process.

While academic programs, particularly PhD paths, place significant emphasis on training students to communicate professionally and effectively explain concepts orally and in writing, to prepare conferences and research and innovation programs, they often neglect the social role young researchers may have during their professional careers. It should be a priority to sensitize young researchers to their role in society and the impact their words can have on the community. Students are taught to consider their audience when shaping the complexity of their speech. They use more complex vocabulary when addressing experts in their field but reduce complexity when presenting their work to non-experts, whether researchers from different fields or the general public. This is important, but these courses should also convey the message that researchers, physicians, and scientists are trusted by many people. When interacting with a non-expert audience, reducing complexity is not the only aspect to consider. Researchers should be fully aware of their social role and influencing power, particularly when dealing with the general public. Before sharing their ideas and opinions, they should consider the impact on the population and communicate in a transparent, coherent, and accurate manner. Personal opinions and doubts about a research topic, when shared within the scientific community, have a controlled effect; while when shared unbalancedly with the population, they can induce fear and rejection, promoting confusion and misinformation [29]. Courses, webinars, and other educational methods should be implemented in universities and companies to increase researchers' and experts' awareness of their social role and the effects of un-adapted communication on people and their perception of innovation.

At the same time, society's awareness should be raised by ensuring easy access to information presented in plain and clear language. People should be sensitized to the implications and benefits of different types of innovation, including the associated ethical issues. Ethics concepts, for example, should be taught from primary school to start sensitizing children to these important topics [28]. Simple ethical concepts could be introduced in games or reading sessions. This will help shape future generations to be aware and more sensitive to ethical themes. Additionally, research entities and governments should adopt a transparent and clear communication style and interact more effectively to create educational events for both young and adult audiences, showcasing what medical research means and entails. This will bridge the existing gap between experts and the general public and foster trust in science by enhancing understanding of the scientific process among people. Collaborations between research institutions and museums, for example, can be implemented to organize interactive exhibitions. Laboratory visits (where feasible in terms of safety) for kids, students, and the general public should be encouraged, focusing not only on the methodology but also contextualizing researchers' work within a broader scenario that includes ethical principles, responsibilities, hopes, and hard work. The objective should be to help people understand that experts in the medical field are first and foremost human beings who have fears like anyone else. Society should be sensitized to the fact that progress is something to be built together, led by experts but requiring effective cooperation among all stakeholders, including the general public. If science (represented by researchers, physicians, doctors, pharmaceutical managers, etc.) does not adopt an open communication approach with society, bridging the communication gap with simpler, transparent, and more participatory messages, society will never fully trust science. It will continue to perceive every new technology as dangerous or reject it completely, potentially compromising the future of health innovation.

To achieve this goal, comprehensive information campaigns to address disinformation and misinformation should be strongly implemented. Researchers, together with governmental agencies, should take the lead on websites, TV channels, social media platforms, etc., to ensure the dissemination of trusted and transparent information that is easily accessible to the general public, protecting them from misleading and fake messages that can undermine people's health and the future of medical innovation.

4 Conclusions

The application of nanotechnology in medicine has the potential to profoundly transform our lives. However, nanomedicine is still in its early stages of social introduction across various medical fields, and its perception and understanding at the societal level are limited. It is crucial to prioritize a clear focus on the ethical and societal implications of nanomedicine to guide its progress. Misleading and biased communication, stemming from a weak ethical consciousness, can significantly impact the social acceptance of medical innovation, and this cannot be overlooked.

This chapter has examined and discussed the roles of ethics and communication in medical innovation. Its aim is to raise awareness among researchers about the importance of their ethical and social responsibilities, as well as the influential power their opinions can have on society. Moving towards the future of medicine requires effective and straightforward communication, respect for human and animal welfare, and robust collaboration among researchers, experts, and government institutions to sensitize society to the scientific and ethical aspects of medical innovation.

References

1. Pautler M, Brenner S (2010) Nanomedicine: promises and challenges for the future of public health. Int J Nanomed 5:803–809
2. Soares S, Sousa J, Pais A, Vitorino C (2018) Nanomedicine: principles, properties, and regulatory issues. Front Chem 6(360)
3. Jiang L, Carmichaels KA (2019) Innovating nanoethics. AMA J Ethics® 21(4):313–316
4. AIFA homepage. https://www.aifa.gov.it/vaccini-covid-19. Last accessed 06 Apr 2023
5. Huang X, Kong N, Zhang X, Cao Y, Langer R, Tao W (2022) The landscape of mRNA nanomedicine. Nat Med 28:2273–2287
6. Kazemi A, Majidinia M, Jamali AA (2014) The question of ethics in nanomedicine. J Clinic Res Bioeth 5:193
7. Resnik DB, Tinkle SS (2007) Ethics in nanomedicine. Nanomedicine (Lond) 2(3):345–350
8. Van Calster G (2006) Regulating nanotechnology in the European Union. Nanotechnol, Low Bus 3(3):359–372
9. Ponce Del Catillo AM (2009) Nanos in the human body—medical perspectives and ethical concerns. HesaMag #01
10. OECD website. https://www.oecd.org/coronavirus/policy-responses/transparency-communication-and-trust-the-role-of-public-communication-in-responding-to-the-wave-of-disinformation-about-the-new-coronavirus-bef7ad6e/. Last accessed 06 Apr 2023
11. Porat T, Nyrup R, Calvo RA, Paudyal P, Ford E (2020) Public health and risk communication during COVID-19—enhancing psychological needs to promote sustainable behavior change. Front Public Health 8:573397
12. Emanuel EJ, Upshur REG, Smith MJ (2022) What Covid has taught the world about ethics. N Engl J Med 387(17):1542–1545
13. Venkatashiva RB, Gupta A (2020) Importance of effective communication during COVID-19 infodemic. J Family Med Prim Care 9:3793–3796
14. Resnik DB (2015) Bioethical issues in providing financial incentives to research participants. Medicoleg Bioeth 5:35–41
15. Central University Research Ethics Committee (CUREC). https://researchsupport.admin.ox.ac.uk/files/bpg05paymentsandincentivesinresearchv10pdf. Last accessed 25 Aug 2023
16. AIRC. https://www.airc.it/cancro/affronta-la-malattia/come-affrontare-la-malattia/studi-clinici. Last accessed 06 Apr 2023
17. NHS Clinical trials homepage. https://www.nhs.uk/conditions/clinical-trials/. Last accessed 06 Apr 2023
18. Informed consent—European Commission. https://ec.europa.eu/research/participants/data/ref/fp7/89807/informed-consent_en.pdf. Last accessed 06 Apr 2023
19. Shah P, Thornton I, Turrin D et al (2022) Informed consent. In: StatPearls [Internet]. StatPearls Publishing, Treasure Island (FL)
20. Petryna A (2007) Clinical trials offshored: on private sector science and public health. BioSocieties 2:21–40
21. NIH website. https://www.nia.nih.gov/health/what-are-clinical-trials-and-stu-ies#:~:text=Clinical%20trials%20are%20research%20studies,safe%20and%20effective%20in%20people. Last accessed 06 Apr 2023
22. Juma C (2016) Innovation and its enemies: why people resist new technologies, 1st edn. Oxford University Press, Oxford; New York
23. Boddice R (2016) Vaccination, fear and historical relevance. Hist Compass 14(2):71–78
24. NIH website. https://www.niehs.nih.gov/research/resources/bioethics/whatis/index.cfm. Last accessed 14 June 2023
25. Steneck NH (2007) Introduction to the responsible conduct of research. Revised edition. https://ori.hhs.gov/sites/default/files/2018-04/rcrintro.pdf
26. Doemeny L, Knerr P (2017) The importance of ethical conduct in scientific research. American Chemical Society

27. Sauer MA, Truelove S, Gerste AK, Limaye RJ (2021) A failure to communicate? How public messaging has strained the COVID-19 response in the United States. Health Secur 19(1):65–74
28. Gülcan NY (2015) Discussing the importance of teaching ethics in education. Soc Behav Sci 174:2622–2625
29. Nelson T, Kagan N, Critchlow C, Hillard A, Hsu A (2020) The danger of misinformation in the COVID-19 crisis. Mo Med 117(6):510–512

Data Quality, Data Diversity and Data Provenance: An Ethical Perspective

Edoardo Ramalli

1 Introduction

In 2006, the British mathematician Clive Humby used for the first time the expression "Data is the new oil." Although at the time this phrase might have seemed almost like a provocation, we know today it is a reality. Since then, there has been a real "data-oil" rush, as we have witnessed terms such as "big data", "data analytics", and "artificial intelligence" entering ubiquitously everyday life. Universities have also adapted to this phenomenon. They started to propose degrees on these topics to meet the growing demand for data-related positions. Companies, sometimes also dazzled by the promising results of employing a large amount of data and also to maintain a certain appeal with the product's market, pushed hard to launch applications of this kind or created job positions that included these buzzwords in the job title.

Artificial intelligence and, more generally, data science-related topics have brought both positive and negative benefits. Indeed, numerous scandals made evident the lack of responsible development [1], creating even stronger social inequalities [2]. As a countermeasure, the public sector is increasingly forced to pursue ethical policies in its algorithms while being kept under constant scrutiny and criticism [2].

Nowadays, the capabilities of artificial intelligence are unquestioned. In some fields and tasks, they are even deemed to overpower human capabilities, although the evaluation of such capacities is still debated [3]. To achieve such levels, there is a need to manage, process, and analyze large amounts of data. Given these preconditions, a data scientist uses the data to create a predictor, i.e., a tool that predicts a situation based on what happened in the past. In these terms, the predictor learns patterns in the data, hence the term "machine learning". As tools, predictors are difficult to interpret, but due to their fast ability to perform tasks, they are gradually supplanting many human tasks, even of social and ethical significance.

E. Ramalli (✉)
Department of Electronics, Information and Bioengineering, Politecnico di Milano, Milan, Italy
e-mail: edoardo.ramalli@polimi.it

S. Chiodo et al. (eds.), *Improving Technology Through Ethics*,
PoliMI SpringerBriefs, https://doi.org/10.1007/978-3-031-52962-7_4

Because of these promising prospects, data scientists are often under pressure to quickly deliver a product such as a predictive model. However, such rapid and thus often uncontrolled development implies that developers do not consider aspects such as ethical implications, which depend on how the product has been used or built [4]. In other words, data scientists often have no idea of their responsibility in developing these products, so they act carelessly and indifferently to these problems and their possible consequences [5]. Other scholars, on the other hand, believe that this technical-social gap is not due to carelessness but rather to the incredible complexity of the problems to be addressed [5]. In her book *Weapons of Math Destruction*, Cathy O'Neil points out that data scientists need to become aware of the fact that big companies exploit their skills to achieve business goals without taking care of the consequences [6]. What is not disputed is that given the obvious problems and the active discussion about what is ethical, there is a regulatory void about these products, while the existing frameworks do not entail troubleshooting documentation for practitioners [1, 2, 5]. It is fundamental to fill this gap, by including in the discussion both ethical and technical perspectives as well as the many practitioners involved in the design, delivery, and employment of a data-driven product [2, 7].

Nowadays, processing very large amounts of data with fewer and fewer resources, including human ones, has become a need. To keep the pace, employing automated tools is necessary where possible. In the meantime, these algorithms are becoming more and more complex, and it is becoming increasingly challenging to explain and interpret results—much more than in the past, when they were designed for simpler tasks. Understanding is being replaced by trusting algorithms (and technology, in general) that can perform human tasks fast, precisely, and reliably.

The development of predictive models is the result of a long pipeline of tasks, where data is processed and is the main protagonist. These tasks vary from data collection and preparation to annotation and visualization. Ethical requirements can be viewed in terms of Data Quality dimensions [8], so that only data that meet ethical requirements are to be considered of high quality. Starting with Data Quality (Sect. 2), this chapter describes, from a technical perspective, two other critical ethical aspects that can be seen as Data Quality dimensions: Data Diversity and Data Provenance. In particular, it follows the design principles for qualitative ethical research in the era of big data [9]. These topics are critical since they can be translated into fairness, neutrality, and transparency in data analysis [10]. Section 3 deals with the problem of diversity and neglected minorities in predictive models. Instead, Sect. 4 presents a solution to practical limitations of Data Quality and Data Diversity. In summary, this chapter discusses the primary source of ethical problems for predictive models from a technical perspective. This discussion will be helpful for any context in which it is needed to process large amounts of data in an ethical manner. The next sections follow this structure: first, the ethical problem is presented with real-world examples, then a practical solution is suggested. The chapter concludes with a final discussion in Sect. 5.

2 Data Quality

Today, data is generated and collected at an unprecedented rate. Data Quality is a fundamental aspect of data management and directly affects everyday life, since many decision-making applications are built on a massive amount of data, also known as Big Data [11–13]. Data Quality has been increasingly investigated in recent years, as it defines several Data Quality dimensions that allow to objectively quantify the quality of a data set by checking if the rules associated with a dimension are fulfilled and how often. The most common Data Quality dimensions are: completeness, consistency, accuracy, uniqueness, and timeliness.

There are many examples of poor Data Quality with ethical implications in the literature. For instance, in 2000, the US Presidential election caused quite a stir due to, among other issues, poor Data Quality of the voter registration database [14]. The data were affected by duplicated registrations (uniqueness), incorrect addresses (accuracy), and missing information (completeness). All these aspects could be identified by a data profiling activity that properly assessed Data Quality dimensions. Similar problems were registered in the US healthcare system [15], supply chain [16] and finance [17], resulting in an estimated cost of around billions of dollars per year [12, 18, 19].

These examples show clearly how better Data Quality implies ethical applications and vice versa. Although the application of the Data Quality rule and its measurement is objective, which rules to define and check and which Data Quality dimension to investigate are not. In fact, the concept of "fitness for use" allows each data scientist to decide which the Data Quality dimensions are of interest to them based on the situation they are dealing with. At the same time, in each particular domain, there is no single way of identifying rules and verifying their criteria.

Table 1 is a toy example of the challenges related to Data Quality. Let us consider completeness as the Data Quality dimension under investigation. Thus, it is more or less arbitrarily decided that completeness is an important Data Quality dimension for the scenario described by the toy example and has to be investigated. In simple terms, completeness measures how many cell data entries are missing in a data set. In other words, if the data set can be represented as a table like in Table 1, the completeness for each column counts the number of empty cells. In this case, the quantification of completeness and its definition are straightforward. However, in practical terms, how to define and apply the rule is not. In Table 1, considering the column "Age", completeness is three out of four or 75%. Regarding "Driving License ID", the completeness should be 50%. If the data is observed more carefully, "Ashlee" is only seven, thus she cannot have a "Driving License ID". Therefore, her corresponding "Driving License ID" cell is not empty. Thus, it is not incomplete. Table 1 reports a simple case to handle, but, in general, the investigation of completeness requires a deep knowledge of the dataset's domain, and a user can define more restrictive rules with respect to others as well as which Data Quality dimension to investigate. To conclude, domain and problem complexity determine subjectivity in the Data

Table 1 A toy example about the arbitrariness of the definition and application of data quality rules

Name	Age	Driving license ID
Tom	22	A1234
Elizabeth		A2312
Ben	25	
Ashlee	7	

Quality check and, thus, together with the assessment, they are fundamental, as they provide additional details about how the assessment is performed.

3 Data Diversity

Data diversity is often under the spotlight in the ethical debate, often paired with a different but related concept: bias. A biased algorithm, predictive model, or dataset cause a high stir since the seriousness of its implications is easily tangible and immediate to understand, also by people without a technical background. In simple terms, bias in data occurs when the entities of a dataset are overrepresented with respect to others. Therefore, when employed for data analysis or to build a data-driven predictive model, adopting a biased dataset results in a prejudiced outcome due to the disproportion of information. Over the last few years, many stories related to the ethical consequences of using biased data have (unfortunately) hit the news. Some examples are: Florida's system for predicting the risk of recidivism [20], the Amazon-AI recruiting system [21], Pennsylvania's tool for screening child welfare [22], and Google Translate's gender bias [23]. These applications are all built on datasets containing bias, mainly based on race or gender. Studies also have shown a "meta-" bias in academia, highlighting how the investigation of possible ethical problems in AI-related fields does not often take into account the misrepresentation of minority groups, such as disabled people [24]. Enhancing diversity in a dataset would bring two kinds of benefits: first, it would mitigate the risk of excluding groups that are considered underrepresented or overrepresented in the ethical debate; and second, it would boost engagement in building more powerful and accurate applications [25].

Moreover, in this case, assessing the heterogeneity (or coverage) of a dataset is a time-consuming task; therefore, it is fundamental to employ a semi-automatic approach that synthetically summarizes whether the ethical requirements are fulfilled. One of the proposed solutions entails the quantification of the database coverage, in which, after identifying the dimension characterizing a domain, a corresponding multidimensional matrix counts the amount of data for each possible combination in the dataset [26]. More precisely, each data entry in the dataset assumes a value for each characteristic or property of the domain. Hence, the matrix dimensions correspond to the identified characteristics of the domain. Each dimension is expected to be able to assume a given set of possible values, which will therefore correspond to the cardinality of the dimension itself. As a result, the final user has a qualitative and quantitative overview of the diversity of the dataset. The solution also proposes a synthetic index summarizing the coverage of the dataset as a ratio between the number of accounted combinations with at least *k* cases over the total number of the existing combinations in a domain where *k* is given. A toy example in Fig. 1 represents the visualization of this concept. The domain is characterized by three dimensions or characteristics: "Age", "Eye Color", and "Height". Each dimension can assume a precise set of possible values. For instance, in the case of "Eye Color", the values are "Blue" and "Green". In this example, the total number of possible combinations is eight. A diverse dataset that fulfills ethical requirements in terms of bias should include at least one element in each box (or bucket, i.e., a combination of property values for each dimension), and, in the general case, it should not include boxes that are over- or underrepresented with respect to others. However, like in the case of Data Quality, there are some drawbacks. The index, depending on how it is used, could result in misleading and unethical results. Considering the example in Fig. 1 with respect to the case of "Eye Color". The multidimensional matrix accounts only for the "Blue" and "Green" colors, but, for example, not for "Brown". Therefore, if the possible values of a domain dimension are not present, or a dimension is not present at all, or the range of a bucket is improperly generous, such as in the case of "Age", there is a risk of having a high diversity score but without having a diverse database in practice. Solutions, like the multidimensional matrix, are fundamental for processing large amounts of data in the Big Data era, but it is also fundamental to properly set dimensions and their possible values. Data Provenance, in the next section, will be fundamental to overcome these limitations. Suppose every scientific and nonscientific community can identify the diversity dimension and their possible values uniquely. In that case, Data Diversity can be measured with this methodology, thus allowing to assess ethical requirements.

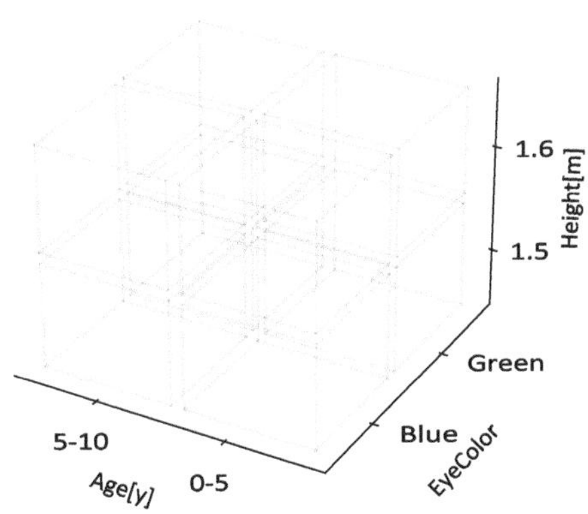

Fig. 1 A visualization of the assessment of the data diversity methodology

4 Data Provenance

Data Quality and Data Diversity are two countermeasures that attempt to reconcile the need to process large amounts of data and to check that ethical requirements are met. However, these approaches, due to how they are defined to digest "Big" Data, generally summarize the results in performance indexes. Unfortunately, due to the domain complexity in which they are applied, there is not a single way to define the indexes and how they are computed. Therefore, these metrics could be subjective and thus responsible for misleading the fulfillment of ethical requirements. Data transparency can mitigate these limitations. In the literature, data transparency is considered not as an ethical principle per se, but rather as a pro-ethical prerequisite to enable other ethical practices or principles [27]. In fact, Data transparency, in practice, consists of providing information about the origin of data and the process to which they have been subjected. The data record containing such information is called Data Provenance or data lineage.

Recently, there has been increasing recognition of the importance of Data Provenance in data-sensitive applications such as artificial intelligence [28]. The employment of data provenance information is fundamental in critical operations where there are several ethical hazards, or it is challenging to ensure trust [29]. Trustworthiness must be shown at several levels in the process of data production and use. It should characterize the raw data, the methodology, the analysts, and the organizations that offer services based on the data set. Data Provenance can be employed to keep track of the adequacy of the process, i.e., the chain of trust [30]. Future users of a service or a data set can gain trust by being allowed to reproduce every single step of data production [31]. In general, Data Provenance is not only helpful in replicating results but also in tracking down errors and accountability [32].

In practical terms, Data provenance is a collection of metadata containing information about entities, activities, and users involved in producing, transforming, or using data. Together with the provenance metadata, the provenance data model must be provided to the data user since it is fundamental for understanding what is stored in the provenance record. In fact, the provenance data model defines the schema of the provenance metadata in a domain by relating the various entities, activities, and people involved in the data creation or in the following data processing stages. According to the literature, the design of the provenance data model should follow the data sheet directives [33], which recommend representing what is strictly necessary for the design of the data preparation pipeline. In fact, a provenance data model can have different levels of granularity [34], and thus different verbosity. In any case, all the information to replicate the steps, from data source to final deployment of the data, must be recorded and provided to the data user. However, like in the case of data quality and data diversity, Data Provenance entails the human element, thus a subjective factor. Data Provenance, hence trustworthiness, depend on the definition of the model that is used to keep track of relevant aspects of the world [30]. In other words, there is no single definition of what is relevant to be reported and what is not.

Thus, we need to trust the analysts or the data providers, their training and accreditation being clearly important [30]. In any case, regardless of what is stored inside, the Data Provenance record enables a stronger reliability of data since "more eyes mean fewer lies, and fewer mistakes" [30].

The W3C PROV data model is often used to design the provenance data model [35]. Figure 2 represents the W3C PROV data model, i.e., a conceptual provenance model built around three elements: Entity, Activity, and Agent [35]. An Entity is something for which we want to trace the provenance. It can be a single entry in a dataset, like an image or the entire dataset, or a single document. An Activity is an operation performed on an entity to produce another entity or another version of it. Typical activities could be the translation of a document into another language or the cropping or tagging of an image. Finally, the Agent is something or someone that bears the responsibility for an action or an entity. An example could be the annotator of a set of images or the data analyst that made a particular decision based on some observations. Each element in the W3C PROV data model is connected to another one by a pre-defined set of relationships as shown in Fig. 2.

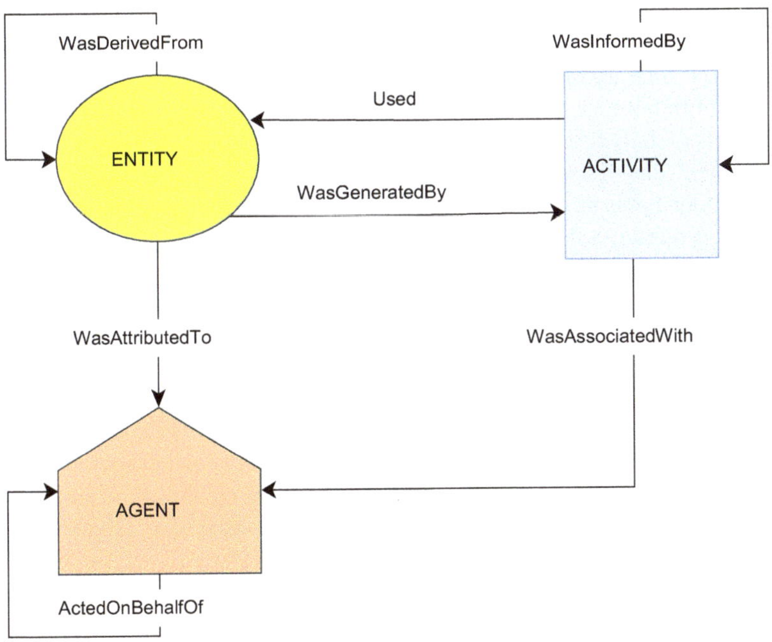

Fig. 2 The W3C PROV data model

5 Discussion and Conclusions

Nowadays, predictive models and, more in general, algorithms that process large amounts of information are fundamental to continuously offer services with limited resources. In particular, AI components bring two benefits: they speed up processes and transfer responsibility. Today, it is possible to develop an intelligent agent to make challenging decisions regarding social aspects of daily life. Even though relying on an algorithm for a health diagnosis or a court judgment could sound scary, people are getting used to the practice of relying on technology to replace human activities and duties. After decades of technological improvements, we have developed an unconscious trust in the technology being able to deliver results faster and more reliably. However, until now, technology has been replacing human activities mainly in repetitive and easy tasks, so that outcomes were easy to understand and verify. Today, algorithms are very complex since they try to mimic human reasoning. Due to such complexity, it is no longer possible to easily verify them, but trust in technology remains the same.

Due to the increasing demand for these promising tools, product designers and developers are often under pressure to deliver them. As a result, there are no resources to carefully analyze the implemented product from an ethical perspective addressing ethical issues such as discrimination or prejudice. These gaps have great media resonance, both because of the implications and because these products are used in many aspects of daily life. Most of the time, the problem is not the algorithm itself but the "corrupted" data fed into it. Recent technologies, such as neural networks, "just" learn the hidden patterns in the data: if the data is biased, the algorithm is biased, too. As society is biased, a dataset representing a particular social activity is probably biased, too. However, when designing and developing an algorithm or a predictive model, developers have the opportunity to measure and mitigate ethical issues. It should be easier than changing unethical behaviors in our society. Sometimes, the opposite happens. The unethical behavior is enlarged when delivering a data-driven product. In both cases, it is fundamental to stick to data management-related ethical issues accounting for Data Quality, Data Diversity, and Data Provenance.

This chapter discusses these data management aspects as mitigation for ethical issues from a technical perspective—although the ongoing ethical debate prefigures many different perspectives. It first presents the problems in terms of examples and then corresponding solutions. The mitigations are introduced keeping in mind that we should address the ethical issues in a way that allows the processing of large amounts of data. On the other hand, providing automatic or semiautomatic tools that do not act unethically is a very challenging task. An ethical debate could be opened about whether it is preferable to use these not totally reliable tools to assess ethical requirements or not to use them at all. As discussed in this chapter, the best we can do is to foster awareness in those who will use a product or data. Data Quality, Diversity, and Provenance are scalable techniques to mitigate technical-ethical issues, but they come with limitations. Data Quality ensures that the "garbage-in garbage-out" paradigm of a data-driven model is mitigated, ensuring quality data for building these tools. However, which Data Quality dimension to investigate and how to apply it remains a subjective choice. Similarly, Data Diversity measures a dataset's heterogeneity to create awareness of the bias in the data and, thus, in the algorithms. As earlier said, it is not always objective on how to assess it. In the end, Data Provenance can overcome the previous limitations. In principle, it records, using a given model, all the actions that happened to an entity. Therefore, it can be used to track all the aspects of the design and creation of a predictive model. Unfortunately, even in this case, simply providing the provenance record is not enough. The provenance data only tracks what is specified in the provenance data model. In this setting, the origin of the ethical problem in the big data era could have a new and different perspective: it is currently missing the engineering share in assembling a long and complex pipeline of a product, such as an algorithm, a predictive model, or an analysis report. Due to other constraints, data scientists often just assemble various components of a data pipeline without reasoning about the drawbacks of such linking. On the other hand, engineers should be characterized by the ability to understand when and how it is safe to connect various parts of a complex system. In this case, engineers, after being properly trained, should consider the provenance record and make conscious ethical decisions in designing their products.

References

1. Jagadish HV et al (2019) The responsibility challenge for data. In: Proceedings of the 2019 international conference on management of data. SIGMOD'19. Association for Computing Machinery, Amsterdam, Netherlands, pp 412–414
2. Stoyanovich J (2019) TransFAT: translating fairness, accountably and transparency into data science practice. In: 1st international workshop on processing information ethically, PIE@ CAiSE 2019
3. Lebovitz S, Levina N, Lifshitz-Assaf H (2021) Is AI ground truth really 'true'? The dangers of training and evaluating AI tools based on experts' know-what. In: The dangers of training and evaluating AI tools based on experts' know-what, pp 1501–1525
4. Saltz JS, Dewar N (2019) Data science ethical considerations: a systematic literature review and proposed project framework. Ethics Inf Technol 21:197–208
5. Barocas S, Boyd D (2017) Engaging the ethics of data science in practice. Commun ACM 60(11):23–25
6. O'Neil C (2017) Weapons of math destruction: how big data increases inequality and threatens democracy. Crown
7. Werthner H et al (2023) Digital humanism: the time is now. Computer 56(1):138–142
8. Firmani D, Tanca L, Torlone R (2019) Ethical dimensions for data quality. J Data Inf Qual (JDIQ) 12(1):1–5
9. Hesse A et al (2019) Qualitative research ethics in the big data era. Am Behav Sci 63(5):560–583
10. Stoyanovich J, Abiteboul S, Miklau G (2016) Data, responsibly: fairness, neutrality and transparency in data analysis. In: International conference on extending database technology
11. Wang RY, Strong DM (1996) Beyond accuracy: what data quality means to data consumers. J Manag Inf Syst 12(4):5–33
12. Redman TC (1998) The impact of poor data quality on the typical enterprise. Commun ACM 41(2):79–82
13. Garcia S, Luengo J, Herrera F (2015) Data preprocessing in data mining. Springer

14. Jones D, Simons B (2012) Broken ballots: will your vote count? CSLI Publications, Stanford
15. Mans RS et al (2015) Data quality issues. In: Process mining in healthcare: evaluating and exploiting operational healthcare processes, pp 79–88
16. Hazen BT et al (2014) Data quality for data science, predictive analytics, and big data in supply chain management: an introduction to the problem and suggestions for research and applications. Int J Prod Econ 154:72–80
17. Du J, Zhou L (2012) Improving financial data quality using ontologies. Decis Support Syst 54(1):76–86
18. Batini C, Scannapieco M (2016) Data and information quality: concepts, methodologies and techniques. Springer International Publishing, Cham
19. Haug A, Zachariassen F, Van Liempd D (2011) The costs of poor data quality. J Ind Eng Manag (JIEM) 4(2):168–193
20. Dressel J, Farid H (2018) The accuracy, fairness, and limits of predicting recidivism. Sci Adv 4(1):eaao5580
21. Dastin J (2018) Amazon scraps secret AI recruiting tool that showed bias against women. In: Ethics of data and analytics. Auerbach Publications, pp 296–299
22. Glaberson SK (2019) Coding over the cracks: predictive analytics and child protection. In: Fordham Urban Law J 46:307
23. Prates MOR, Avelar PH, Lamb LC (2020) Assessing gender bias in machine translation: a case study with google translate. Neural Comput Appl 32:6363–6381
24. Lillywhite A, Wolbring G (2019) Coverage of ethics within the artificial intelligence and machine learning academic literature: the case of disabled people. Assist Technol
25. Drosou M et al (2017) Diversity in big data: a review. Big Data 5(2):73–84
26. Ramalli E, Pernici B (2021) Know your experiments: interpreting categories of experimental data and their coverage. In: CEUR workshop proceedings, vol 2929, pp 27–33
27. Turilli M, Floridi L (2009) The ethics of information transparency. Ethics Inf Technol 11:105–112
28. Werder K, Ramesh B, Zhang R (2022) Establishing data provenance for responsible artificial intelligence systems. ACM Trans Manag Inf Syst (TMIS) 13(2):1–23
29. Tullis JA, Bandana K (2021) Where is the provenance? Ethical replicability and reproducibility in GIScience and its critical applications. Ann Am Assoc Geogr 111(5):1318–1328
30. Hand DJ (2018) Aspects of data ethics in a changing world: where are we now? Big Data 6(3):176–190
31. Cranmer K et al (2015) Analysis preservation in ATLAS. J Phys: Conf Ser 664(3):032013
32. Herschel M, Diestelkamper R, Ben Lahmar H (2017) A survey on provenance: what for? What form? What from? VLDB J 26(6):881–906
33. Gebru T et al (2021) Datasheets for datasets. Commun ACM 64(12):86–92
34. Chapman A et al (2020) Capturing and querying fine-grained provenance of preprocessing pipelines in data science. Proc VLDB Endow 14(4):507–520
35. Belhajjame K et al (2013) PROV-DM: the PROV data model. W3C Recomm 14:15–16

Ethics-Aware Application of Digital Technologies in the Construction Industry

Ania Khodabakhshian

1 Industry 4.0 Revolution in the Construction Industry

The construction industry is complex and unique in nature, consisting of multiple interconnected activities and decisions with direct safety, social, ecological, and economic impact on human lives and cities. While the traditional and manual design, engineering, and construction methods were identified as labor-intensive, time-consuming, inefficient, uncollaborative, and unsafe [1], the Industry 4.0 revolution and advancements of digital technologies started revolutionizing and digitalizing the construction industry, automating repetitive tasks, increasing project's performance, optimizing the supply chain, simulating and controlling the physical processes, proactively mitigating risks, enhancing safety, and optimizing decision [2]. The digital transformation of the industry is happening due to (a) generation and documentation of enormous real-time data by novel technologies like the Internet of Things (IoT), Sensors and Laser scanners, Cloud Computing, Big Data Analytics, Text Mining, etc., (b) structured documentation of construction data in interoperable and collaborative cyber-physical systems like Digital Twins, Building Information Modelling (BIM) models, Project Information Systems, etc. for knowledge transfer, and (c) broad application of Artificial Intelligence (AI) to learn relationships between input and output data records, extracting insights and detecting patterns from historical data, and generalizing the rules to future projects to predict values and automate decision-making processes [3–5]. Figure 1 presents the major digital technologies used in the construction industry.

However, as these technologies are used by architects, engineers, and project managers to create living and work spaces for humans, they should be human-centric and socially harmless to them and the environment they live in. Hence, the

A. Khodabakhshian (✉)
Department of Architecture, Built Environment and Construction Engineering, Politecnico di Milano, Milan, Italy
e-mail: ania.khodabakhshian@polimi.it

S. Chiodo et al. (eds.), *Improving Technology Through Ethics*,
PoliMI SpringerBriefs, https://doi.org/10.1007/978-3-031-52962-7_5

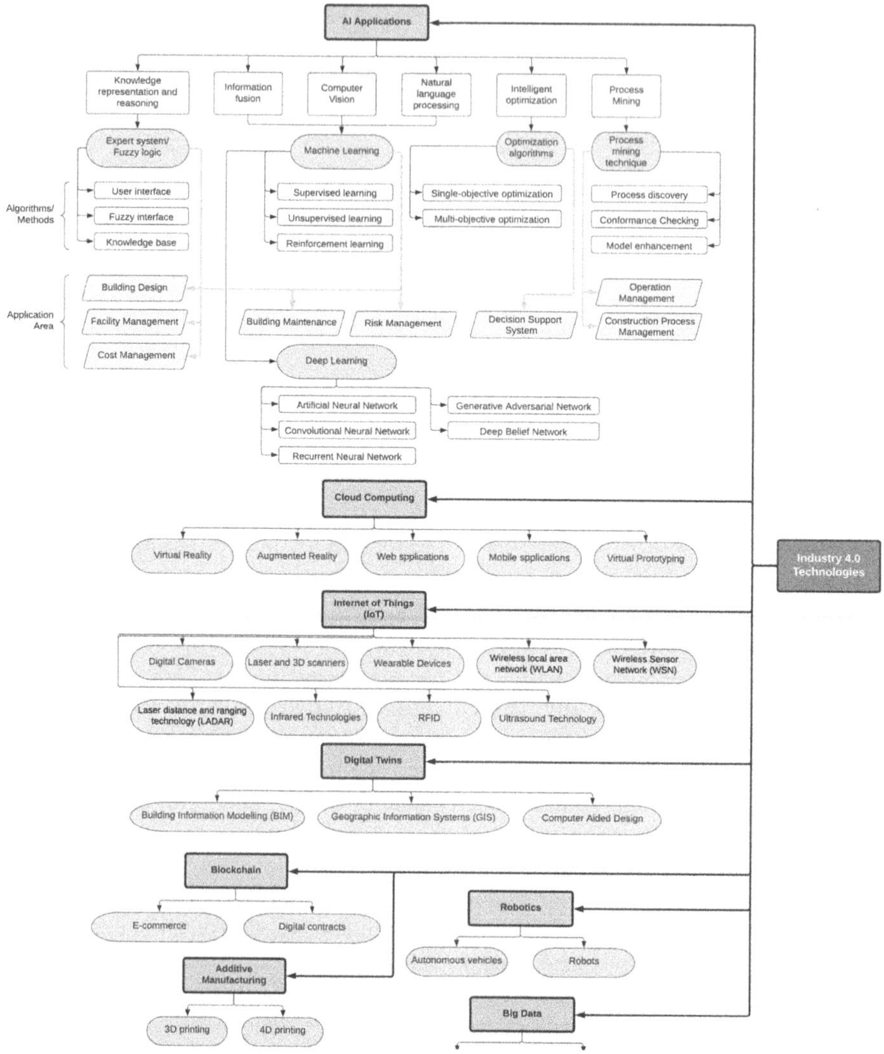

Fig. 1 Industry 4.0 technologies in the construction industry [1, 6–9]

social, ethical, and moral responsibilities of such technologies, beyond the technical standards, should be addressed as prerequisites to implement them in the design, construction, and operation phases of projects [10]. This initiative gives rise to the concept of industry 5.0 for a transition toward a sustainable, human-centric and resilient industry that looks beyond productivity and efficiency as end goals [11, 12]. However, the tradeoff between delivering building projects faster, cheaper, and at higher quality with following ethical and moral regulations is not always easy, considering the great capital, lengthy process, numerous stakeholders, and complex data environment in construction projects.

The inherent complexity, bureaucracy, and the change-resistant culture in the construction industry, as well as technology switching costs and market uncertainty, slow down the industrial evolution and digital-data-driven systems integration in projects, which can be perceived both as a threat and an opportunity [13, 14]. The lack of established standards and references on data security and protection, knowledge transfer, interoperable formats, data ownership, digital contracts, etc., as well as the lack of proper training to familiarize the professionals with digital technologies, are hindering the widespread application of industry 4.0 technologies in practice [2]. However, it leaves room to integrate ethic codes and standards in writing such standards for a human-centric and ethic-aware application of technology in the sector. This chapter aims to determine the key aspects and components of an ethics-aware technology application framework in the construction industry, with a focus on AI, as the closest technology to human cognitive processes of learning, reasoning, perception, planning, and self-correcting [3].

2 Ethics of Digital Technologies

There are many drivers and benefits in using digital technologies on a broad scale that interests construction companies including (a) lower business costs of manpower with reduced need for human intervention, (b) real-time collection, processing, and monitoring of data compared to human agents, (c) faster prediction of risks, delays, and clashes, due to simulating the digital version of the project in advance, (d) enhancing workers' safety by using robots in dangerous activities, and (e) allocating machinery and resources to the activities optimizely [15]. AI specifically aims to boost labor efficiency by 40% and double annual economic growth rates by 2035 [16].

Nevertheless, to obtain these benefits, companies have to bear high implementation costs, immensely change their established process and organizational structures, secure data privacy and cyber security, and educate their personnel [2]. The implementation of AI models may raise various ethical, moral, and social dilemmas that digital agents are not equipped to resolve. Ethical dilemmas occur when a decision must be made between two alternatives that are both ethically problematic and can result in unethical behaviors, not following some values, norms, and laws in the society like justice and inclusion [17]. For instance, when a choice needs to be made between a safer road design, and a road design that preserves more nature. On the other hand, a social dilemma happens when there is a conflict between self and collective interests [18]. For instance, if a generative design tool is used to optimize only one perspective of the building, such as the mechanical and electrical systems or the architectural design, it may not be the best for the entire project and raise conflicts of interests.

In order to resolve these dilemmas, first, a description of ethics and moral principles is required. Morality refers to internal principles and codes of conduct regarding right and wrong put forward and accepted by a group or an individual, upheld by the individuals themselves, not by any law or regulation [19]. On the other hand, and in a broader sense, ethics is a set of principles and disciplines society conveys to guide individuals on how to act in different circumstances based on what is deemed right or wrong in that given situation, considering personal development of oneself and loved ones [20, 21]. Both principles are influenced by various cultural, social, organizational, geographic, contextual, and historical factors and are intended to promote positive interactions and prevent negative ones [21], and to create stronger, safer, healthier, more inclusive, and more just societies.

Applied ethics defines protective boundaries alongside the autonomy and authority given to a human agent or a technology agent, like AI. AI ethics, as a set of values, principles, and techniques employing widely accepted standards of right and wrong in such dilemmas [22], is developed to address ethical issues related to the AI system and its developers [23] that may arise when designing and developing AI (e.g., human biases that exist in data, data privacy, and transparency), and ethical issues caused by AI (e.g., unemployment, wealth distribution, and change in power dynamic).

As evident in the theoretical definition, AI ethics is directly connected to human-technology relationships and trust. Trust is a psychological state comprising the trustor's intention to accept vulnerability based on positive expectations of the intentions or behavior of the trustee [21]. Trust can motivate an action or collaboration, in this context, public engagement in AI vast applications. Once the society or a part of the society, in this case, the construction industry professionals, are convinced that AI systems are human-centered and serve humankind, are reliable, are compatible with the established workflows and reinforce best practices instead of replacing them, are interactable and controllable and are inclusive and accessible to all groups of society, they will be more open to accept and promote them [24].

A better knowledge of human rights laws can help AI algorithm developers eliminate or at least alleviate the potential harms, biases, discrimination, and invasion of privacy issues in AI [23]. Therefore, it is important for the AI developers, as well as the researchers and practitioners who are going to use it, to understand existing ethical principles that need to be applied to construct ethical AI models following the same principles. A better knowledge of human rights laws can help AI designers and engineers eliminate or at least alleviate the discrimination and invasion of privacy issues in AI [23]. Such AI models capable of translating ethical principles into practical and measurable metrics throughout the planning, development, deployment, tracing, and operation phases of the system are addressed as moral autonomous systems, ethical controllers [25], artificially intelligent autonomous systems [26], or responsible AI [25, 26] in the literature. The main ethical principles that need to be addressed in developing ethical AI models for construction projects are mentioned in the following subsections [17].

2.1 Sources of Data, Biases, and Discriminations

The accuracy of AI algorithms depends on the quality and quantity of input data. In the construction industry, with inherent data scarcity and subjectivity issues, ensuring objectivity, completeness, generalizability, fairness, diversity, and confidentiality of collected data becomes even more critical. For instance, if the data is gathered through surveys and interviews, the interviewees should be representatives of all the gender/age/ethnic minority groups for the data not be discriminated [10], or if any wearable devices are used to measure the stress level and health parameters of the workers, they should also include female workers' parameters even if they are proportionally much less. Gender and race bias reflected in Natural Language Processing (NLP) studies and text mining is another example of human bias that needs to be addressed, with the gender binary being challenged by social and liberal groups, and gender overshadowing other variables like race, age and occupational role for classification, regardless of actual importance [27].

Moreover, the collected data should not be focused only on one dimension, but all the political, economic, social, technological, environmental and legal dimensions and ethical codes that apply to a real situation [28]. AI-driven technologies reflect and amplify patterns of marginalization, inequality, and discrimination in the social context [22], which are critical issues to consider when choosing the input data and algorithm to use for a specific problem. For instance, AI-assisted hiring technologies may increase the unemployment of a specific group due to community-based gender and racial occupational segregation [29]. In the male-dominated construction industry, with the continuous decrease in skilled workmanship, women can play a key role in managing digital innovations. Therefore, to avoid gender bias, discrimination, and filter bubbles in developing digital technologies, diversity represents the key, urging to fundamentally rethink technological progress, enabling new business models in construction, and keeping up with digitization in a sustainable, unbiased, and inclusive manner [10].

2.2 Trust in AI Technology and Decisions

Trust between the AI agents and human counterparts is determined by (a) the legal environment and standards, and (b) the technical aspects of the system, warranting safety, security, transparency, and explainability [24]. All trust drivers, like belief in the adequacy and competence of current ethics codes, standards, and laws, the perceived impact of AI on jobs, familiarity and understanding of AI, and AI's impact on society fall under one of these two categories [30]. One of the most controversial topics is whether AI will create more jobs or eliminate some jobs [30]. While construction workers are being replaced by robots or architects by generative design algorithms, AI is changing the jobs and work dynamic, creating new job opportunities like machine learning engineers or data scientists. Aiming to assist humans and improve their lives, AI is automation and optimization of burdensome, repetitive, and dangerous tasks like construction field activities, but on the other hand, it can cause huge job replacement and unemployment, which puts human rights protection under the spotlight. Promoting proper education and training on AI can help the labor force keep up with the changes in jobs and benefit from the capabilities of AI [31].

Another issue to consider is the level of trust and reliance on decisions and predictions made by AI. Neither our knowledge nor AI's knowledge, which is learned from the data we provide it, is perfect and constantly updated. Therefore, it is questionable whether predictions made by AI are reliable and accurate enough to base decisions on, and what the optimum level of authority is to hand over to AI. While AI assessments are more objective, the task they perform uses the belief system and biases of their creators, perpetuating subjective behaviors based on outdated belief systems [17]. A simple solution seems to be encoding rules and regulations in the algorithms or digital twins; however, AI cannot still question the underlying ethics or biases in the inserted rules or the social context. Overtrust and overreliance on AI could demotivate the human agents to question algorithms' prediction or take responsibility for the final result [17]. Based on a thematic analysis in Architecture, Engineering, and Construction (AEC) literature conducted by Emaminejad and Akhavan (2022) [32], four key dimensions to build trust in the AI algorithms are (a) explainability and interpretability, (b) reliability and safety, (c) performance and robustness, and (d) privacy and security; which will be analyzed in the next subsections.

2.3 *Explainability and Interpretability*

Explainability and interpretability are intertwined topics in AI and robotics, indicating the state where the operations of a system can be understood by a human through introspection or explanation [33]. In the construction industry, the explainable AI (XAI) is especially important since the professionals are not familiar with advanced structures of AI algorithms like black-box; therefore, neither they can understand the workflow between the input and output (i.e., explainability), nor the output would be meaningful to them (i.e., interpretability). XAI aims to transform black-box to white-box AI, allowing users to intuitively interpret how algorithms and models arrive at an output [34], and keeping them in their decision loop. XAI helps build trust and ensure model accuracy, fairness, and transparency [35].

While Artificial Neural Networks (ANNs) learn and solve problems without being programmed and with the minimum of human guidance in a full black-box manner [36], Probabilistic Graphical Models (PGM) like Bayesian Networks and Structural Equation Modeling (SEM) and Fuzzy Logic are some closer examples of XAI that can be easily comprehended by the user in a graphical mode using a symbolic representation of domain knowledge and pre-defined rules rather than complicated algorithms or statistics [9, 37]. However, there are still challenges to be addressed in XAI, including a tradeoff between accuracy and interpretability and explaining AI competencies versus explaining decisions to users [35].

2.4 Reliability and Safety

Reliability and safety indicate the capacity of the models to avoid failures or malfunctions and exhibit the same expected behavior over time [38]. The real threat of AI is not taking control over the world but the undetected errors, failures, and misuses resulting from wrong and unexplained decisions. Therefore, as important components of trust, both concepts are related to trust from a performance (rather than a moral) angle and find meaning in human-technology interactions [39]. The safety and reliability concepts can be seen from both technical and cognitive/psychological (e.g., affect) perspectives. As a related dimension, and in the context of adopting AI and robotics in AEC, safety also has to be studied from both technical (i.e., avoiding accidents) and psychological (i.e., job replacement) viewpoints. However, overtrust and overreliance on technology may lead to ignorance toward risks associated with trusting the technology; the solution to which is training and educating the human users about the abilities, reliability, and failure modes of the AI agent to avoid overtrust and undertrust [40].

2.5 Performance and Robustness

Performance and robustness have a close relationship with ability, accuracy, or competence concepts [41]. Performance indicators play a key role in trust trajectories by offering standardized and common criteria to compare different AI models [42]. Robustness refers to performance consistency in different situations where the AI model or the robotic system is deployed, which in construction translates to the ability to successfully transfer models between different projects, sectors, and locations considering the uniqueness and context-drivenness of projects without compromising the user's trust in the system [30]. These are important issues in the construction industry as the environment and workflows are ever-changing within and between projects, and consistency of performance is critical to project success [24]. While almost all performance measures in construction ultimately target a metric related to the cost or schedule of the project [9, 41], factors such as quality, safety records, sustainability measures, productivity metrics, and inspection outcomes are as important as ultimate cost and schedule. For instance, simulating construction site layout planning in BIM can significantly increase the construction process performance and efficiency.

2.6 Privacy and Security

Privacy and security are pillars of ethical or responsible AI, where the protection of human identity and sensitive data, as well as the protection of the system against attacks that breach privacy, are addressed [43]. Privacy is defined as the right not to have personal data acquired, observed, and used. Security in AI is ensuring the confidentiality of data, preserving information integrity, and guaranteeing immediate data availability when desired. In construction projects, the privacy of workers' sensitive data collected from wearable devices or sensors [44], as well as cybersecurity issues related to the application of cameras or drones, are critical challenges [11, 44]. Data regarding contract information, transactions, blueprints, photos, and project personnel are considered confidential and careful attention must be paid to guarantee data confidentiality and prevent data leaks [24]. This issue becomes more critical in cloud-based and virtual applications and databases, where the model ownership and liability need to be specifically defined and access to external bodies need to be limited. Ignorance of such sensitive rights will hinder the application of robotics and autonomous systems for data collection and storage.

2.7 Autonomy and Accountability

AI is a big umbrella term that includes various technologies, applications, types, and subfields, which can be categorized into three groups based on their level of autonomy and authority [6]: (a) Artificial Narrow Intelligence (ANI), (b) Artificial General (or "strong") Intelligence (AGI), and (c) Artificial Super-Intelligence (ASI). ANI aims to automate some repetitive and learnable activities without the ambition to substitute human intelligence or authority, or decision-making. AGI aims to match human-level intelligence in any field and type of human activity and is capable of complex decision-making. ASI aims to exceed human intelligence and faculties, staying unbeatable by any human mind [45]. The current state of AI application in the AEC industry is mostly the ANI and, to some extent, the AGI. However, with moving toward the ASI, the system's autonomy and trustworthiness can become disputable.

AI has four types of application in construction given its level of automation [46]: (a) Automated Intelligence, i.e., the automation of manual and cognitive tasks without involving innovation, (b) Assisted Intelligence, i.e., efficient and accurate tracking, analysis, and surveillance of data [47], (c) Augmented Intelligence, i.e. algorithmic decision-making systems to foster the construction personnel in their activities and decisions, and (d) Autonomous Intelligence, i.e., decision support systems without human interaction or robots that can replace the workforce completely [48].

The stronger the AI system autonomy, the more accountability it should bear and the more serious the accountability becomes. For instance, there are many regulations and contract laws for BIM model ownership, liability, and copyright issues. Such regulations should be composed in the AI algorithms application domain as well to lower the risks and issues derived from data confidentiality and ownership. Determining who is responsible for wrong predictions resulting in problems and accidents, whether it be the programmer, data owner, or end-user, is difficult. In cases where an AI decision results in harm to a person, there is an accountability gap, as it is unclear whether to hold the AI itself responsible or the developers or project managers/engineers who used it. To address this issue, it is necessary to consider the different levels of AI autonomy and their relationship with human control [22].

3 Ethical Standards, Laws, and Regulations

Today, standards, laws, and regulations have strategic importance in the application of digital technologies, providing advantages in the research, development, and operation phases by offering guidelines to change some parts or connecting different software modules in a reliable way by standardized protocols [49]. Unlike ethics and moral, which are perceived as internal and mental beliefs or expectations from other people, laws and standards are generated by a group of people to set administrative limitations or recommendations, translating the philosophical beliefs into practical protocols.

Codes of ethics, regulations, and standards developed by governments, companies, technical organizations, trade unions, and societies are the most influential factor in building public trust toward AI [31, 50]. Such regulations enforce compliance with ethical and moral values and determine human responsibility in the development and deployment of intelligent systems, filling the gap that emerges from the increased automation of decisions [51].

AI standards have been developed on national and international levels [31, 50], such as the AI policy observatory principles developed by the Organization for Economic Co-operation and Development (OECD) [31, 50], compromising over 620 national AI policies from over 60 countries, Institute of Electrical and Electronics Engineers (IEEE) standardization Association's "Global Initiative on Ethics of Autonomous and Intelligent Systems" project on AI standard [52], Future of Life Institute [53], International organizations for standardization such as IEC and ISO standards [54], European Union, European Commission's High-Level Expert Group on AI (European Commission, 2019) [55], US National Institute of Standards and Technology (NIST) [56], and many more. These standards and regulations need to be supported by AI community, State administration, and Standardization bodies [57].

There are two approaches for developing ethical standards [58]:

(a) The Consequentialist justification evaluates and weights actions in terms of their long-term impact and outcome, their implied costs and benefits.
(b) The Deontological approach suggests sticking to some basic principles that embody the notion of values in a society and directly evaluating the rightness or wrongness of actions and policies without considering their direct impact explicitly.

It is important to find a balance between the two principles to resolve ethical dilemmas. Departing from general principles, considered value judgments about specific situations and outcomes and iterative adjustments will be made in both principles till a consolidation is found. These principles aim to promote (a) inclusive growth, sustainable development, and well-being through trustworthy AI, (b) human-centered values and fairness throughout the AI system lifecycle, (c) transparency, explainability, and responsible disclosure regarding AI systems, (d) robustness, security, and safety, and systematic risk management, (e) accountability of AI systems [59], (g) culture of awareness and alignment of AI with ethical values, (h) the competitiveness of the construction industry, and (i) education and awareness-raising for the potentials, opportunities and risks of AI [10].

4 Ethical AI Framework

Transforming AI ethics and standards into a practical framework applicable to construction projects is not easy. Various components and processes need to be considered and assessed in terms of compliance with both the ethical standards and the enterprise processes in order to guarantee the system's capability of making ethical and moral decisions. There are various ways to educate AI systems and transform them into ethical agents, including (a) implicit ethical agents, when a machine's actions are constrained to avoid unethical outcomes, which could significantly limit AI application, (b) explicit ethical agents, stating explicitly allowed and forbidden actions, and (c) full ethical agents, when machines have consciousness and free will like humans and can make their own decisions, although it is still in the research phase [60].

However, ethics metrics are not easy to measure like hard performance metrics, such as safety and reliability; hence they are overlooked by companies. In the construction industry, the project budget and quality still come first, even if the ethical laws have been disrupted. Therefore, companies face a tradeoff between focusing on AI advancement to realize profit maximization or focusing on AI ethics to ensure societal benefits derived from AI [23]. Another challenge to face is the shift of risks in the construction industry from known space, like financial or delay risks, to unknown space, like the liability of contractors if an accident was cause by a robot or a wrong design generated by AI [34].

4.1 Ethics-Aware Design and Implementation of AI

Ethics must be reflected in different layers of AI algorithms development and implementation. Design engineers must be fully aware of the ethical challenges, biases, cyber security risks, etc., and inspect the potential flaws in the system design [61]. The AI system itself should be able to reflect on the ethical and social principles and standards embedded by developers to make socially significant decisions and avoid unethical behaviors [62]. And finally, in order to implement a responsible and ethics-aware AI system, the ethical mindset and culture in the organization and among employees need to be promoted. In order to develop and implement an ethics-aware AI system, fairness should be addressed at different levels: (a) Data Fairness, when the input data is representative, relevant, accurate, and generalizable to the entire study sample, (b) Design Fairness, when used target variables, features, processes, or analytical structures are not unreasonable, morally objectionable, or unjustifiable, (c) Outcome Fairness, when the outcomes are not discriminatory or inequitable impacting on the lives of the people, and (d) Implementation Fairness, when users are sufficiently trained to implement them responsibly and without bias [22].

4.2 Integration of AI in Project Team and Processes

The project team requires motivation to accept AI in their engineering and management processes and change the usual way of doing their activities. Autonomous motivation is based on the fulfillment of the three psychological needs: autonomy, competence, and relatedness [63], all of which will be impacted by AI. The consequences of trusting AI in decisions, hindering the possibility for the team to learn from their mistakes, limiting the productive conflicts from which the team grows and strengthens their relationships, and the new team dynamics and collaboration models when applying and relying on AI are some critical concerns and motivation hinders. If we let AI decide for us, what does this do to our motivation and the motivation and performance of a project team [64]? On the other hand, AI has the potential to take over all the repetitive and standard work and free up humans to deal with creative thinking, which could be a positive motivation for the project team. Therefore, there is a dilemma between the motivation and performance of the project team when AI comes into the picture.

4.3 Education and Training on AI

Education and training have been introduced as the most straightforward solution to solve the ethical, moral, and social issues of technology application in the industry by previous researchers [22, 25]. Education has different target populations: (a) AI applications developers need to be educated about ethics codes and standards, (b) the AI algorithms need to be educated in embedding these standards, (c) management-level policymakers need to be educated on the pros and cons and ethical considerations on AI, and (d) end users, in this case, engineers, architects, and project managers, need to be educated with basic knowledge and hands-on experience on AI algorithms and codes of ethics. This way, a bilateral human-centric relationship will be shaped between the technology and human agents, and the change-resistant culture of the industry will change, giving more exposure and space for AI implementation on a vast scope.

Figure 2 presents a holistic list of harms, biases, and challenges in three steps of AI application in construction companies, i.e., Database collection, AI model development, and Implementation for Decision Making, that need to be considered and addressed for an ethical and responsible AI application.

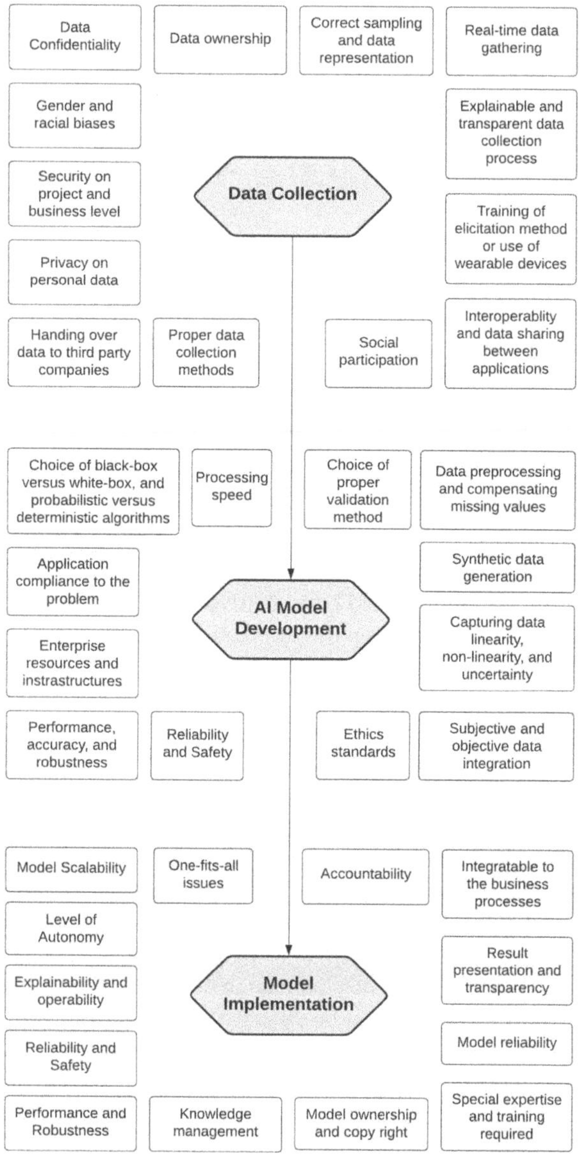

Fig. 2 Ethics aware AI framework components in construction projects [22, 24, 57]

5 Conclusion

Industry 4.0 technologies and data-driven digital systems are digitalizing the construction industry, proposing new ways of doing repetitive, burdensome, time-consuming, and inefficient tasks in projects. While these technologies have the potential to save great capital and resources, increase safety and productivity, and enhance collaboration in construction projects, the proper infrastructures and standards for their ethical, harmless, and fair application is not yet developed. Ethics in digital technologies, especially AI, is a niche and novel area that is being more emphasized, moving toward the Industry 5.0 revolution. The lack of established AI standards in the construction industry can be perceived as a threat but also as an opportunity and driver to develop such practical standards with an ethical point of view, enabling machines to somehow reason and make ethical and moral decisions, which leads to no harm to society. This chapter delineated the main ethics-aware technology implementation factors that need to be addressed when composing ethical standards and guidelines and developing technology implementation frameworks. The limitation of the study was the lack of similar studies in the construction realm, and most AI ethics literature was conducted in the manufacturing, IT, or healthcare sectors. Therefore, it is difficult to compare the findings of this study with a similar benchmark. As a potential future study, the proposed framework can be implemented on real-world projects and in construction companies to highlight the practical challenges and potential solutions offered by codes of ethics.

References

1. Regona M, Yigitcanlar T, Xia B (2022) Opportunities and adoption challenges of AI in the construction industry: a PRISMA review. J Open Innov Technol Mark Complex 8. https://doi.org/10.3390/joitmc8010045
2. Oesterreich TD, Teuteberg F (2016) Computers in industry understanding the implications of digitisation and automation in the context of Industry 4.0: a triangulation approach and elements of a research agenda for the construction industry. Comput Ind 83:121–139. https://doi.org/10.1016/j.compind.2016.09.006
3. Darko A, Chan APC, Adabre MA et al (2020) Artificial intelligence in the AEC industry: scientometric analysis and visualization of research activities. Autom Constr 112:103081. https://doi.org/10.1016/j.autcon.2020.103081
4. Forcael E, Ferrari I, Opazo-vega A (2020) Construction 4.0: a literature review. Sustainability 12. https://doi.org/10.3390/su12229755
5. Bilal M, Oyedele LO, Qadir J et al (2016) Big Data in the construction industry: a review of present status, opportunities, and future trends. Adv Eng Inform 30:500–521. https://doi.org/10.1016/j.aei.2016.07.001
6. Abioye SO, Oyedele LO, Akanbi L et al (2021) Artificial intelligence in the construction industry: a review of present status, opportunities and future challenges. J Build Eng 44:103299. https://doi.org/10.1016/j.jobe.2021.103299
7. Chowdhury T, Zealand N, Adafin J et al (2019) Review of digital technologies to improve productivity of New Zealand construction industry. J Inf Technol Constr 24:569–587. https://doi.org/10.36680/j.itcon.2019.032

8. Akinosho TD, Oyedele LO, Bilal M et al (2020) Deep learning in the construction industry: a review of present status and future innovations. J Build Eng 32:101827. https://doi.org/10.1016/j.jobe.2020.101827
9. Pan Y, Zhang L (2021) Roles of artificial intelligence in construction engineering and management: a critical review and future trends. Autom Constr 122:103517. https://doi.org/10.1016/j.autcon.2020.103517
10. Weber-Lewerenz B (2021) Corporate digital responsibility (CDR) in construction engineering—ethical guidelines for the application of digital transformation and artificial intelligence (AI) in user practice. SN Appl Sci 3. https://doi.org/10.1007/s42452-021-04776-1
11. Kozlovska M, Klosova D, Strukova Z (2021) Impact of Industry 4.0 platform on the formation of Construction 4.0 concept: a literature review. Sustainability 13. https://doi.org/10.3390/su13052683
12. European-Commission, Directorate-General for Research and Innovation, Breque M, De Nul L, Petridis A (2021) Industry 5.0: towards a sustainable, human-centric and resilient European industry. European Commission Publications Office, Luxembourg
13. Zhang L et al (2021) Artificial intelligence in construction engineering and management. Springer, Singapore
14. Manuel J, Delgado D, Oyedele L et al (2019) Robotics and automated systems in construction: understanding industry-specific challenges for adoption. J Build Eng 26:100868. https://doi.org/10.1016/j.jobe.2019.100868
15. Yaseen ZM, Ali ZH, Salih SQ, Al-ansari N (2020) Prediction of risk delay in construction projects using a hybrid artificial intelligence model. Sustainability 12:1–14. https://doi.org/10.3390/su12041514
16. Purdy M, Daugherty P (2016) Why artificial intelligence is the future of growth. In: Crawford K, Whittaker M (eds) Remarks at AI Now: the social and economic implications of artificial intelligence technologies in the near term. AI Now Institute, pp 1–72
17. Arroyo P, Schöttle A, Christensen R (2021) The ethical and social dilemma of AI uses in the construction industry. In: Alarcon LF, González VA (eds) Proc. 29th annual conference of the International Group for Lean Construction (IGLC29), Lima, Peru, pp 227–236. https://doi.org/10.24928/2021/0188
18. Van Lange PAM, Joireman J, Parks CD, Van Dijk E (2013) The psychology of social dilemmas: a review. Organ Behav Hum Decis Process 120:125–141. https://doi.org/10.1016/j.obhdp.2012.11.003.236
19. Stanford Encyclopedia of Philosophy (2020) The definition of morality. https://plato.stanford.edu/entries/morality-definition/. Accessed 4 Apr 2023
20. Stanford Encyclopedia of Philosophy (2022) Moral theory. https://plato.stanford.edu/entries/moral-theory/#:~:text=Ethics%20is%20generally%20understood%20to,exemplify%20human%20excellence%20of%20character
21. Kuipers B (2020) Perspectives on ethics of AI: computer science. In: Dubber MD, Pasquale F, Das S (eds) The Oxford handbook of ethics of AI. Oxford University Press, Oxford, UK
22. Leslie D (2019) Understanding artificial intelligence ethics and safety systems in the public sector. The Alan Turing Institute
23. Siau K, Wang W (2020) Artificial intelligence (AI) ethics: ethics of AI and ethical AI. J Database Manag 31:74–87. https://doi.org/10.4018/JDM.2020040105
24. Emaminejad N, Akhavian R (2022) Trustworthy AI and robotics: implications for the AEC industry. Autom Constr 139:104298. https://doi.org/10.1016/j.autcon.2022.104298
25. Trentesaux D, Karnouskos S (2022) Engineering ethical behaviors in autonomous industrial cyber-physical human systems. Cogn Technol Work 24:113–126. https://doi.org/10.1007/s10111-020-00657-6
26. Russell BS (2016) Should we fear supersmart robots? Scientific American, pp 58–59
27. Koolen C, van Cranenburgh A (2017) These are not the stereotypes you are looking for: bias and fairness in authorial gender attribution. In: Proceedings of the first workshop on ethics in natural language processing, Valencia, Spain, pp 12–22

28. Flores F (2012) Conversations for action and collected essays: instilling a culture of commitment in working relationships. CreateSpace Independent Publishing Platform
29. Altman M, Wood A, Vayena E (2018) A harm-reduction framework for algorithmic fairness. IEEE Secur Priv 16:34–45. https://doi.org/10.1109/MSP.2018.2701149
30. Gillespie N, Lockey S, Curtis C (2021) Trust in artificial intelligence: a five country study. The University of Queensland and KPMG Australia
31. Galindo L, Perset K, Sheeka F (2021) An overview of national AI strategies and policies. Going Digital Toolkit Note
32. Emaminejad N, North AM, Akhavian R (2021) Trust in AI and implications for the AEC research: a literature analysis. In: ASCE international conference on computing in civil engineering (i3CE). Florida, USA
33. Haibe-Kains B et al (2020) Transparency and reproducibility in artificial intelligence. Nature 586:E14–E16. https://doi.org/10.1038/s41586-020-2766-y
34. Pillai VS, Matus KJM (2020) Towards a responsible integration of artificial intelligence technology in the construction sector. Sci Public Policy 47:689–704. https://doi.org/10.1093/scipol/scaa073
35. Gunning D, Stefik M, Choi J et al (2019) XAI-explainable artificial intelligence. Sci Robot 4:4–6. https://doi.org/10.1126/scirobotics.aay7120
36. Mcaleenan P (2020) Moral responsibility and action in the use of artificial intelligence in construction. Proc Inst Civ Eng – Manag Procure Law 173:166–174. https://doi.org/10.1680/jmapl.19.00056
37. Hvam L, Mortensen NH, Riis J (2008) Knowledge representation and forms of reasoning for expert systems. Product customization. Springer, Berlin, Heidelberg, pp 197–217
38. Hoff KA, Bashir M (2015) Trust in automation: integrating empirical evidence on factors that influence trust. Hum Factors 57:407–434. https://doi.org/10.1177/0018720814547570
39. Malle BF, Ullman D (2021) A multidimensional conception and measure of human-robot trust. In: Trust in human-robot interaction. Elsevier, pp 3–25
40. Merritt SM, Heimbaugh H, LaChapell J, Lee D (2015) I trust it, but I don't know why: effects of implicit attitudes toward automation on trust in an automated system. Hum Factors 55:520–534. https://doi.org/10.1177/0018720812465081
41. Hoffman RR, Johnson M, Bradshaw JM, Underbrinck A (2013) Trust in automation. IEEE Intell Syst 28:84–88. https://doi.org/10.1109/MIS.2013.24
42. Glikson E, Woolley AW (2020) Human trust in artificial intelligence: review of empirical research. Acad Manag Ann 14:627–660. https://doi.org/10.5465/annals.2018.0057
43. Orr W, Davis JL (2020) Attributions of ethical responsibility by artificial intelligence practitioners. Inf, Commun Soc 23:719–735. https://doi.org/10.1080/1369118X.2020.1713842
44. Choi B, Hwang S, Lee S (2017) What drives construction workers' acceptance of wearable technologies in the workplace?: indoor localization and wearable health devices for occupational safety and health. Autom Constr 84:31–41. https://doi.org/10.1016/j.autcon.2017.08.005
45. Müller VC, Zalta E (2020) Ethics of artificial intelligence and robotics. The Stanford encyclopedia of philosophy, Winter 202. Metaphysics Research Lab, Stanford University, Stanford, CA
46. PWC (2017) The essential eight technologies board byte: augmented and virtual reality
47. Golparvar-Fard M, Peña-Mora F, Savarese S (2015) Automated progress monitoring using unordered daily construction photographs and IFC-based building information models. J Comput Civ Eng 29. https://doi.org/10.1061/(ASCE)CP.1943-5487.0000205
48. Feng C, Xiao Y, Willette A, McGee W et al (2015) Vision guided autonomous robotic assembly and as-built scanning on unstructured construction sites. Autom Constr 59:128–138. https://doi.org/10.1016/j.autcon.2015.06.002
49. Zachariadis I (2019) Standards and the digitalisation of EU industry economic implications and policy service
50. OECD (2019) AI policies and initiatives. Artificial intelligence in society. Paris. https://doi.org/10.1787/cf3f3be0-en

51. Theodorou A, Dignum V (2020) Towards ethical and socio-legal governance in AI. Nat Mach Intell 2:10–12. https://doi.org/10.1038/s42256-019-0136-y
52. IEEE (2021) IEEE standard model process for addressing ethical concerns during system design, IEEE, 7000-2021
53. Future of Life Institute (2017) Asilomar AI principles. Future of Life Institute. https://futureoflife.org/ai-principles/?%0Acn-reloaded=1
54. ISO (2021) Artificial intelligence. ISO/IEC JTC 1/SC 42. https://www.iso.org/committee/6794475/x/catalogue/p/1/u/0/w/0/d/0
55. The-European-Commission (2019) Ethics guidelines for trustworthy AI. https://ec.europa.eu/digitalsingle-%0Amarket/en/news/ethics-guidelines-trustworthy-ai
56. NIST (2021) AI standards. https://www.nist.gov/topics/artificial-intelligence/ai-standards
57. Mezgár I, Váncza J (2022) From ethics to standards—a path via responsible AI to cyber-physical production systems. Annu Rev Control 53:391–404. https://doi.org/10.1016/j.arcontrol.2022.04.002
58. Nagel T (2021) Types of intuition—intimations of morality. London Rev Books 43(11):1–11. https://www.lrb.co.uk/the-paper/v43/n11/thomas-nagel/types-of-intuition
59. Yueng K (2020) Introductory note to recommendation of the council (OECD). Int Leg Mater 59:27–34. https://doi.org/10.1017/ilm.2020.5
60. Anderson M, Anderson SL (2007) Machine ethics: creating an ethical intelligent agent. AI Mag 28:15–26. https://doi.org/10.1609/aimag.v28i4.2065
61. Bostrom N, Yudkowsky E (2014) The ethics of artificial intelligence. In: Frankish K, Ramsey W (eds) The Cambridge handbook of artificial intelligence, 1st edn. Cambridge University Press. Cambridge, pp 316–334. https://doi.org/10.1017/CBO9781139046855.020
62. Wallach W, Allen C (2009) Moral machines: teaching robots right from wrong. Oxford University Press, New York
63. Deci EL, Olafsen AH, Ryan RM (2017) Self-determination theory in work organizations: the state of a science. Annu Rev Organ Psychol Organ Behav 4:19–43. https://doi.org/10.1146/annurev-orgpsych-032516-113108
64. Schöttle A (2020) What drives our project teams? In: Proc. 28th annual conference of the Intern. Group for Lean Constr. (IGLC), Berkeley, California, pp 313–324

Drawing Together: Technology Development for Public Service

Carlos Sandoval Olascoaga, Lillian Chin, Josefina Correa Menendez, Reid Kovacs, Calvin Zhong, and Nicholas de Monchaux

1 Introduction

Cities and buildings are created through collaboration—between designers, and the communities and clients they serve. However, authorship in architecture and urban planning practice is still primarily in the hands of a small and select group of practitioners. Centering this design practice around those whose everyday lives are shaped

C. Sandoval Olascoaga (✉)
School of Architecture and Khoury College of Computer Sciences, Northeastern University, Boston, MA, USA
e-mail: c.sandovalolascoaga@northeastern.edu

Department of Architecture, Massachusetts Institute of Technology, Cambridge, MA, USA

L. Chin
Department of Electrical and Computer Engineering, UT Austin, Austin, TX, USA
e-mail: ltchin@utexas.edu

Department of Electrical Engineering and Computer Science, Massachusetts Institute of Technology, Cambridge, MA, USA

J. Correa Menendez
Brain and Cognitive Sciences, Massachusetts Institute of Technology, Cambridge, MA, USA
e-mail: jcorream@mit.edu

R. Kovacs
Electrical Engineering and Computer Science, Massachusetts Institute of Technology, Cambridge, MA, USA
e-mail: rkovacs@mit.edu

C. Zhong · N. de Monchaux
Department of Architecture, Massachusetts Institute of Technology, Cambridge, MA, USA
e-mail: czhong@mit.edu

N. de Monchaux
e-mail: ndm@mit.edu

S. Chiodo et al. (eds.), *Improving Technology Through Ethics*,
PoliMI SpringerBriefs, https://doi.org/10.1007/978-3-031-52962-7_6

by the design of the built environment is critical to create more ethical cities and buildings, especially to right previous systemic injustices.

One of the biggest impediments to effective community participation in the design of the built environment is the political and social complexity of community participation. Existing participatory design practice often fails to be inclusive and diverse enough to represent community interests, especially in a scalable way. Furthermore, the analog practice of involving communities is often disconnected from the complex digital tools shaping the physical design of the environment. Without a new approach to participation and co-creation for planning and design, the enormous amount of building necessary to respond to current climate challenges will potentially continue to reinforce the inequity that characterize the making of cities today.

We believe that computing can play a role in returning us to a more emergent, and community driven approach to architecture and urban planning. As an interdisciplinary team of computer scientists, urban planners, and architects, we see the biggest potential for computing in the built environment not as a decider of what our cities should look like, but as a tool to amplify the voices of those whose everyday lives are shaped by those designs. By decentralizing the decision-making process, computational design tools and digital technologies have the potential to provide a formal framework for collaboration between architects, data, and communities.

To achieve this scale of collaboration, new kinds of computational tools for participation are needed that can engage our communities and their experiences in the design and planning process. Historically, design and planning tools like Geospatial Information Systems (GIS) or Computer-Aided Design (CAD) software have been developed with the interests of other disciplines in mind, such as natural resource management or the military [1]. This has led existing CAD and GIS tools to favor a central author and interest, to be siloed from each other, and to favor precision over design exploration. These tools' philosophies run counter to the nature of design and planning, which are highly interactive endeavors that evolve through conversation, feedback, and iteration. For computationally mediated community participation to be effective in a design and planning context, we must involve the community in the design and development of the software tools themselves.

In this chapter, we discuss the limitations of existing computational tools for design and planning and our work in developing new tools with historically underserved communities. We first explore the history of early computer mapping to demonstrate that the motivations and implementations of software architectures can have irrevocable influences on how the built environment is designed by architects and planners. With these historical lessons in mind, the second section of this chapter underscores the need for creating design and planning technologies that are domain specific, accessible, and contextual. We implement this framework through a partnership with Green City Force (GCF), an AmeriCorps program that trains young adults from NYC Public Housing in environmental service. We describe how our close engagement with GCF's staff and AmeriCorps (Corps) members, enabled us to identify critical operational and design challenges that the organization is currently facing. Based on these challenges, we describe a set of modular technological tools that we developed in collaboration with the organization in service of their day-to-day planning and operations. Finally, we show the relevance of these tools beyond this organizational context, and we underscore the need for co-designing planning technologies in collaboration with the end user to ensure their relevance, accessibility, and ethical adoption in practice.

2 Design and Planning Origins of Geospatial Technologies

An understanding of the cultural and computational history of our design tools, technologies, and of our spatial data is critical to implementing these technologies in an ethical manner for urban design and planning. Understanding who these technologies serve, represent, and impact empower us to reframe how urban data, public service, computing, and design can be employed to allow the broadest set of community actors to shape their own environments equitably and holistically.

In this section, we focus on the historical development of GIS, a computer mapping and spatial analysis tool generally employed in design and planning practice. This serves as a case study of how we can use the history of computational tools to think about the future potentials of computing in the built environment. This history is particularly important given that we are at a moment when planners and technologists are again turning optimistically to digital systems to design and plan the built environment. We strongly believe that this history of GIS can provide lessons in which urban planning and software development practices can aspire to build inclusive and open tools and methods through direct collaboration with practitioners and the communities affected by their tools.

We first discuss the history of GIS' precursor, SYMAP, a computer mapping program created in the early 1960s by a practicing architect to expand the access of mapping to designers and planners. Although designed to improve the accessibility of computer maps in design, these tools ultimately resulted in unsuccessful experiments in systems-based design in the 1960s and 1970s. Later, in the early 1980s, computer mapping software was generalized by enterprises to serve larger industries, like oil and gas or the military. While this transition broadened the use of computer mapping and helped establish a multibillion-dollar industry, these software interfaces were less compatible with designers and planners' day-to-day practice. From this historical retrospective, we reiterate our emphasis on using technology to democratize the decision-making process and to amplify planning as a collaborative and visual practice.

2.1 *Computer Mapping for Spatial Management*

In the mid-twentieth century, early programs for computer mapping were developed to automate the analysis and production of cartographic information [2]. Two of the earliest computer mapping programs, the Canada GIS (CGIS) and the University of Washington Computer Mapping Program, were created to facilitate arithmetic calculations with spatial information to optimize spatial decision-making in a rational and quantitative manner [3].

Both CGIS and the University of Washington Computer Mapping Program were disengaged from traditional cartographic and creative design practices. Both programs focused on optimizing the ways of quantifying and cataloguing what existed, instead of providing tools and graphic tools to invent what there could be. These initial attempts to digitize cartographic practices would need more fluent user interfaces to be useful for a creative context.

2.2 SYMAP: Computer Mapping for Urban Planning and Design

To improve the accessibility of computer mapping in a planning and design context, Howard Fisher, an architect by training, developed SYMAP, a computer mapping program for use in mainframe computers.

With a background in architecture, Fisher was less interested in information processing and more interested in producing intuitive ways to represent and communicate information graphically [4]. Fisher's background motivated him to improve the aesthetics and graphic display of spatially distributed data of existing computer mapping programs [5]. This focus on a visual interface would broaden the use of the software in architecture and planning, even by someone with no computer experience or without mathematics beyond high school algebra [6].

Between 1963 and 1964, Fisher, with the programming assistance of Betty Benson, developed SYMAP, short for "Synagraphic Mapping", a neologism for "Drawing Together". This name emphasized the collaboration between the computer and the designer. SYMAP intended to combine the computer's strength in compiling and processing quantitative information with the design input, subjective evaluation of graphic information, and feedback provided by designers [7].

The universality of SYMAP's graphic maps enabled architects, landscape architects, and planners to collaborate in the analysis and design process. Fisher believed that digital mapping had the potential to become a bridge between designers, the public, and stakeholders. For Fisher, the process of digital map-making could have its most profound impact not in resource quantification, but as a graphic representation that could be used for interdisciplinary collaboration in the design of the built environment.

Following SYMAP's initial development at Northwestern University, Fisher was able to secure funding from the Ford Foundation for the subsequent development of SYMAP, and support from Harvard's Graduate School of Design to provide an academic home for Fisher's research enterprise, where he founded the Laboratory for Computer Graphics (LCG) in 1965. Establishing the LCG at the GSD not only provided Fisher with an academic home for his software development enterprise, but it helped him attract talented students and researchers. These early computer mapping pioneers went on to establish the computer mapping corporations like ESRI that have transformed computer mapping software ever since.

However, the initial support for the LCG from government institutions and private foundations aimed to support the use of computer mapping as a tool to improve planning policy through rational decision-making methods. These grants were meant to lead to the application of quantitative decision-making frameworks, such as Systems Analysis, to design and city planning. "Only numbers could lead to rational planning decisions," read Fisher's grant proposal to the Ford Foundation [8]. By the end of the 1960s, the emphasis of the LCG gradually shifted to introduce the techniques of simulation and systems analysis, "with graphic display assuming a supplementary but nonetheless crucial role as a communication medium between data and the human mind" [9].

2.3 ESRI: Computer Mapping as Enterprise Software

"*Really, Now, Who or What is ESRI?*" [10] wrote Fisher, as his work at the intersection of architecture and computer mapping was being superseded by that of his protégé, Jack Dangermond, founder of Environmental Systems Research Institute (ESRI). Unlike Fisher, the aspirations of Dangermond and his company had little to do with design culture, and rather aligned with the technocratic goals of precision and control in the built environment. Ever since its establishment in 1969, ESRI has played a key role in the expansion of GIS to a variety of fields and has had a major impact in shaping our GIS tools and methodologies.

Following his time at Harvard, Dangermond implemented the LCG's tools and practices to establish an environmental consulting business to work initially with city governments and design firms. However, by 1973 it was clear to the academic community that ESRI's goals for computer mapping went beyond planning and design. Originally self-identified as an 'Institute,' and developed under a brief period of non-profit legal status from 1969 to 1973, ESRI's largest impact has been as a multi-billion-dollar business across fields other than design and planning.

ESRI's operations quickly grew into a business to support any project that could benefit from mapping automation. By the mid-1970s, most of ESRI's consulting work was related to Land Capability/Suitability Analysis [11]. Whether the clients were urban designers looking for the sites with the best views and accessibility, or local governments developing the least environmentally impactful land uses, ESRI used a replicable pipeline for planning that was enabled by SYMAP, and other computer programs developed in-house.

ESRI's design and planning process was abstracted into a rational and procedural flow diagram, in which analytical operations were connected to each other by information flowing through them, which was inspired by the techniques of systems analysis. These planning methodologies and graphic diagrams translated the client's goals and requirements into a deterministic database query to select geographic areas, which has ever since become the main way in which users engage with GIS software.

By 1982, ESRI's client list included public works departments that needed automated plans and engineering drawings; oil and gas companies that automated land assessments; foresters that automated environmental overlays for land suitability/capability; and tax assessors that automated municipal tabular files of land related data [12]. To support this specific set of users and scale up its software business, ESRI narrowed down the focus of its computer mapping program to instead provide a robust, well-documented and general use program. ESRI built a generic system that standardized a range of spatial representations into a single information model based on a topological representation of Points, Lines, and Polygons, which ESRI's Professional Services division applied for customers [13]. Professional Services' job was to work with the customers to frame and redefine the problem properly according to ESRI's software offerings. This interaction between both organizations enabled an interplay between the needs and desires of the customer and the computational and time constraints of the software development team.

ESRI's generalizable information model traded abstraction to support a broad set of clients at the price of expressiveness and flexibility, a decision still visible in current GIS programs. This data representation was successful in representing and managing aggregated environmental populations, discrete geographic features, and abstracted government regulations. However, this representation failed to represent the rich and complex qualities of the built environment for use in everyday design and planning practice. The needs of the planning and design community were too subtle, complex, and difficult to generalize into a computer mapping tool targeted to a broader market and were not essential enough to ESRI's business to be part of GIS software specifications.

As general-use GIS systems have continued to be pushed beyond information management, and into everyday urban planning and design practice, the adoption of GIS among planning practitioners has been slow, and not without friction [14]. Due to the historical development of GIS, designers and planners have not been considered the primary audience, resulting in a lack of robust, visual, and graphic-based GIS frameworks for designers and planners.

3 Our Approach: Civic Technologies Through Collaborative Planning

As the above case study into the history of computer mapping has shown us, it is necessary to build design and planning tools through a close engagement with real-world needs to ensure their adoption and accessibility by practitioners. General technologies are insufficient for fully capturing the iterative and collaborative nature of designers working together *with* communities.

To re-center technologies for planning use, we return to Fisher's initial idea of SYMAP, especially the layman version: "Drawing Together." Rather than viewing technology as something limited to only the jargon-filled spaces of academia, our goal is to make our technological interventions truly accessible to the people. We argue that computing can have a major impact in the built environment—as a tool to bring together architects, landscape architects, planners, and the communities that they serve in the design of the built environment. As our software and workflows expand to support a grass-roots approach to thinking about design and the nature of cities, we believe that our software architectures should allow us to follow this model.

3.1 Collaborative Planning History and Theory

Core to our new approach to technology is collaborative planning. We argue that many of GIS' pitfalls as a planning technology follow from its use of rational planning as an architectural basis. Rational planning ideals from the mid-twentieth century that examine the implementation of planning policy and action often overlook identity, power imbalances, and the diverse epistemological worlds that multiple publics, planners, and technologists occupy. Even when public participation is involved in the planning process with historically marginalized groups, without the understanding of the "cultural specific norms, values, and ways of interpreting the world", practices from even the most well-intentioned planners often fall short.

This shortcoming is reflected in Sherry Arnstein's seminal article, "A Ladder of Citizen Participation," which defines levels of participation that range from harmful and patronizing, to truly empowering [15]. At the lower rungs of the ladder of participation, inclusion acts only as therapy or as information; in its extremes, it serves as harmful manipulation of communities. However, at the highest rungs, and to have true participation and citizen power, power is redistributed and delegated amongst the communities in question so that they may be directly involved in decision making. To combat the disconnect between project implementation by planners, practices, planning technologists and the real, multicultural, and diverse communities that the projects affect, it is important to understand and respect the different places we all come from, especially when values, vocabularies, and ways of doing differ. The issues that planners face and plan for in community-based work are not self-evident; instead, the goals, processes, and intents of the project must be intentionally aligned.

Bent Flyvbjerg suggests that implementing planning praxis requires more than just the "analytical, scientific knowledge (*episteme*) and technical knowledge or know-how (*techne*)," and instead, suggests "phronesis," or a kind of judgement based on other factors [16]. Phronesis urges us to move beyond simple conceptions of technology deployment, but to also consider technology beyond its technical components. In other words, how technology is developed and deployed matters. Constraints of usability and cost may be analytical, but the deployment and usage of these technologies in service of participatory planning ideals requires us to grapple just as much with more subjective ideas and political and institutional barriers. This is particularly important because technology often reinforces the bureaucratic and technocratic rhetoric that enhances, rather than challenges, existing power structures.

Given the inadequacies of previous planning technology and the importance of critical technological development, we choose to build a grassroots approach to technology use and deployment. This allowed us to address the barriers to success of a small nonprofit group that has technological needs, while supporting the larger goals of leveraging technology with critical minds towards equity and redirecting the conventional paradigm of technocratic and bureaucratic planning practice in the United States. We implemented smaller scale interventions and technologies that we believe will have a greater impact than a single, technically, and politically complex intervention. This mode of flexible practice has demanded from us an attentiveness to the variety of cultures, experiences, and feelings of its deeply different participants, strategic alignments with (or the lack-thereof) government institutions and policies, and for an openness towards alternate processes and outcomes.

4 Case Study: Drawing Together with Green City Force

We are currently implementing our vision of technology built from collaborative planning principles with our partnership with Green City Force (GCF), a non-profit that works with NYC public housing residents for social, economic, and environmental justice. This project is inspired by SYMAP's history, with the goal to enable community members to "Draw Together" to collaboratively expand the social and ecological impact that community spaces can have across New York City Housing Authority (NYCHA). Rather than previous technologies that were designed for large corporations in mind, *Drawing Together* centers community participation in design and planning and demonstrates how computing could be used to scale community spaces and workforce development programs.

Scoping specific interventions from these goals initially proved challenging due to the large number of stakeholders, the difficulty of synthesizing reparative action with traditional technological approaches, and the need to balance immediate needs with long-term improvements. In the following sections, we describe the context of New York public housing and GCF, outline how we and GCF built our working relationship, defined our joint projects, and the status of those projects.

4.1 Background

NYCHA is New York City's agency for public housing. It houses more than 400,000 residents, in over 170,000 apartments, in over 300 developments, serving mostly Black and Brown New Yorkers. Legacies of racism, displacement, and unjust planning practices have resulted in significant hardships and barriers to social mobility for NYCHA residents. Recent lawsuits, at the local and federal level, have been filed to address some of the squalid conditions and states of disrepair that residents have lived with [17]. In response, NYCHA and the local city government, led first by Mayor Bill De Blasio and now Mayor Eric Adams, have pledged billions of dollars for improvement. These plans include plans for capital improvement and repairs, and a new master plan for NYCHA's open spaces. Under these conditions, and in collaboration with NYCHA, GCF, a nonprofit organization, has built Eco-Hubs in six of the NYCHA developments across all five boroughs in New York.

GCF's Eco-Hubs are a multi-site infrastructure situated in public housing developments that aligns local green services for food, water, waste, and energy behavior change and neighborhood transformation strategically with local, city, state, and national goals for climate and equity. These Eco-Hubs, colloquially known as "farms at NYCHA", are a platform for workforce development, grounded by physical farms. Specifically, GCF hosts an AmeriCorps program for 18–24-year-old young adults living in public housing developments, training them on service, sustainability, and justice.

These farms engage Black and Hispanic youth and promote their capability for self-determination. This is in line with reparative planning, which aims towards "the rejection and dismantling of white supremacy such that life chances become independent of one's ascribed social location" [18]. Rashad Akeem Williams proposes a reckoning that acknowledges the material harms of white supremacy, racial liberalism, and racial capitalism in the United States. That is, to counteract and rebuild from a historically racialized planning agenda, the path forward in planning must bring about material, symbolic, and relational forms of repair. Moreover, this paradigm of planning calls for a "more deeply politicize "therapeutic planning" such that it orients toward the goal of reparative justice," in which participation and community engagement reaches beyond degrees of tokenism into degrees of citizen power [19, 20].

The Eco-Hubs have proven outcomes: more than 500,000 service hours, over 600 graduates, and a graduate job placement rate of over 80%, drastically improving future outcomes and average unemployment rate of their target demographic. GCF's goal is to expand the number of farms, increase their number of graduates, and the social and environmental outcomes of the Eco-Hubs to the scale of NYCHA. However, despite the impact and influence of GCF, they have a principal need to improve outreach and recruiting efforts to bring more residents on as AmeriCorps members. Although GCF has the capacity for more than double the number of cohort members, they have not successfully captured their desired audience and have been unable to fill all available positions.

4.2 Building Our Relationship with GCF

Initially, our goal as MIT researchers was to leverage our technological background to the fullest. Given our past work developing CAD/GIS technologies for collaborative planning, we sought to create a new digital platform for a large-scale community-led design process for the site-selection of Eco-Hubs.

Our initial idea was a web platform that would provide a high-level analysis of GCF's goals, metrics, and community engagement at all stages of Eco-Hub development. We envisioned a single website that would allow GCF to, first, pick a location from all possible NYCHA developments, second, co-design an Eco-Hub within a selected site, and third, improve engagement within an already built Eco-Hub. Each of these design scales would have distinct tools for GCF staff, Corps members and residents to collaboratively draw on maps and discuss what they wanted to see in their community (Fig. 1).

Through the collaborative planning, however, it became clear that these large-scale tools were a mismatch from what GCF leadership wanted. GCF wanted to maintain full design control over the site selection process and were uncomfortable with augmenting their process with data and community feedback since their current methods were serving them well. Rather than insist on presenting a single monolithic technological solution, we shifted our approach to working closely with GCF to identify their challenges and asks. This conversation led us to instead build multiple smaller technologies in parallel. This fostered greater trust between us and GCF as they saw that we understood and respected their primary goal of improving recruiting.

One approach that was particularly appreciated was partnering with GCF in a semester-long practicum course taught at MIT with students from the Architecture and City Planning departments. The course was an extension of the goal setting conversations that our team was having with GCF and resulted in short-term proposals to some of the challenges identified by the students, such as app design or user conversion analysis. We believe that by approaching GCF's concerns from many different design perspectives, GCF leadership was able to select individual approaches that most resonated with them, allowing us to get more structured feedback and better understand their priorities. This trust helped them to take a risk with our longer-term projects, where we explore more technically complex interventions.

Our current approaches include digitizing GCF's existing data streams for easier data collection and creating a model that visualized and summarized the multiple criteria GCF used for site selection. While our current projects do not directly involve CAD/GIS techniques, we argue that these simpler technological solutions are needed to meet GCF and the communities they serve where they are. Scaffolding technology in small steps helps keep communities directly involved in their development, with potential for co-ownership rather than a planner/user relationship, a relationship consistent with best reparative planning practices.

4.3 Digitizing Eco-Hubs

One organizational challenge that GCF faces is data consolidation. Regularly understaffed, members of GCF management have employed non-uniform data collection techniques, specific to each member of the team. This has led to disparate data storage

Fig. 1 Initial wireframe of a collaborative co-design website for selecting and co-designing Eco-Hub sites

locations and formats, such as paper forms, online and offline spreadsheets, and customer relationship management software. While the different data formats work for the distributed functions of GCF, it makes it difficult to gather cross-department statistics. Recruiting, alumni engagement, and fundraising all work separately, but creating a coherent narrative about farm performance and corps members outcomes is difficult with the existing data practices.

To address this, we sought to introduce small-scale modular technology that could supplement GCF's current processes to automate data collection from people and farm operations. These records include information about farm usage, crop yield, recruitment profiles, and GCF-conducted NYCHA audits. This data, once consolidated, could provide extra insights for GCF on their current practices while reducing the burden of data entry. Additionally, a streamlined information storage technique opens the opportunity for additional inputs such as resident feedback, online engagement analytics, data collection from sensors, and automated information aggregation. We aimed to make these technologies accessible for novices, laying the groundwork for potential tech education efforts around the maintenance of these systems and the analysis of the resulting data.

One technology we developed is automated sensor modules to collect real-time and hyperlocal environmental information. These modules collect data that can help GCF members and leadership observe in real time several factors that may impact their farms and measure the outcomes of their work through a quantitative framework. Additionally, the collected data can serve as a platform for fundraising and recruitment by presenting compelling statistics backed by reliable information.

The modules are deployed within the raised beds of the farms and are fully autonomous. They are powered via solar, collect information with onboard sensors, and transmit data with onboard communications. The modules collect information about the soil, ambient air, and weather conditions to monitor factors like soil moisture, concentration of pollutants in the ambient air, and intensity of sunlight at a given location. These can be interpreted to generate insights such as the frequency of crop watering, ideal crops for measured sun conditions, and long-term air quality observations. Additionally, the measured values from the sensor modules can be used to predict the results of larger-scale investments. For example, monitoring the output of the small solar panel on the sensor module can provide an estimate of total energy that could be collected for larger solar panels.

These modules can augment and improve the existing network of sensors in New York. At the time of writing, there are a range of networked environmental sensors throughout the city. Some of these projects do not provide resulting data, while others offer an open network and a consumer sensor. Sensors, such as Plant Spike, require interaction from a smartphone, and do not provide publicly available datasets [21]. Other platforms, such as PurpleAir, offer a consumer air quality measurement system that can be integrated into a larger, public network of air quality measurements, but at a much greater cost ($200–250 per sensor) [22]. Lastly, the Environmental Protection Agency offers a dataset of air quality data collected in various locations around the United States but is sparse and insufficient in a city as dense as NYC [23]. Our sensor module addresses some of these challenges by offering a fully open system that automatically transmits data over a long-range communication channel. This project offers an opportunity to build a platform that can consolidate the publicly available datasets of these other projects.

Finally, the on-site assembly and maintenance of the sensors provides new educational opportunities for the Corps members. Since the modules are assembled with off-the-shelf components, they are simpler and more accessible to service for GCF members at a fraction of the cost of commercially available options. We are in the process of developing an embedded programming and data analysis curriculum that can be used in conjunction with these sensors (Fig. 2).

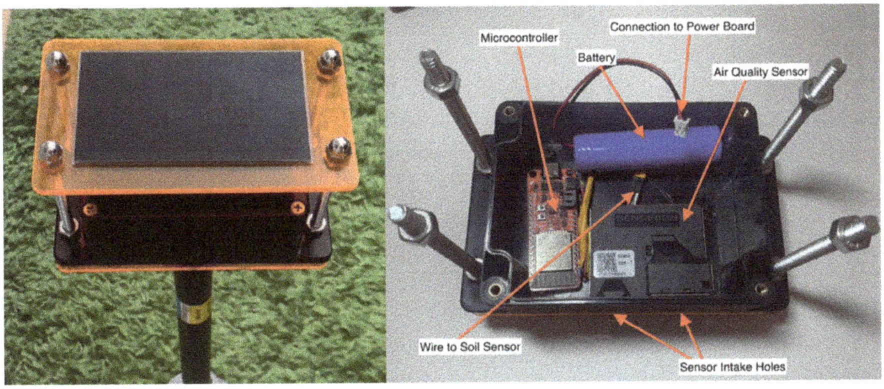

Fig. 2 The sensor module enclosure (left) and labeled internal components (right)

We view these sensor modules as a step to introducing other digitization techniques for existing GCF workflows. As GCF staff become more familiar with supporting technical projects, they may be open to introducing other technical projects for Corps members to work on and support GCF operations. One future project we are considering is a portable optical scanner for paper forms. This system could automatically scan and aggregate data, reducing the time to consolidate and enter data while simultaneously maintaining paper as an accessible interface for less technically literate NYCHA residents.

4.4 Site Allocation Framework

GCF farms have the potential to increase food access in food-scarce regions, increase resilience to flooding in sites with a high flooding risk, and help secure a job upon program completion [24]. The combination of these impacts, as well as sociopolitical concerns, makes Eco-Hub site selection a difficult process for GCF. GCF's current site selection process mostly relies on a time-consuming unstructured process, which includes personal visits to potential NYCHA locations, overlaying maps on transparencies or recommendations from residents.

An optimization framework could provide several advantages to the current analog process. First, it provides a quantitative assessment of the utility of a candidate site. Second, sites that don't meet requirements can be excluded through constraints. Third, we can find the best lower bound on the proportion of individuals of any demographic groups across an arbitrary spatial range, as opposed to performing a manual search after thresholding. Fourth, summary statistics for a candidate set of optimal sites can be readily obtained, facilitating communication of the measurable outcomes that Eco-Hubs can have in NYC. Fifth, the extent to which a GCF staff member values one component of the model's utility over another can be explicitly incorporated through well-defined parameters. In summary, using optimization can provide quantitative tools to help satisfy GCF's qualitative needs.

Conversations with GCF leadership revealed that choosing an optimal site requires considering several factors, including (1) available farmland, (2) the distance that GCF members would travel to work at a given site, (3) population demographics, (4) flooding risk, (5) financial resources. We incorporated these site selection constraints by modeling the utility of a site as a linear combination of resilience to flooding, recruitment potential, employment, and outreach. We identified as key decision variables whether an Eco Hub is built at NYCHA site j, termed $\mathbf{Z_j}$, and whether GCF Corps members travel from NYCHA site i to work at Eco Hub j, termed $\mathbf{X_{ij}}$. The choice of these decision variables stems from the need to identify optimal site locations, subject to several constraints. The binary matrix X allows for excluding sites that would require residents to travel large distances, for example. GCF leadership is also interested in building sites with a large population of individuals within certain age groups. For example, teenagers in the 14–18-year-old range could potentially be recruited to join the program, and people above 60 years old would benefit the most from having fresh produce nearby. Thus, we can aggregate populations in a vicinity of each candidate site and find sites such that the lower bound on the proportion of individuals within each demographic group is maximized. This is known as the max–min framework, an approach that has been used broadly in optimization problems and discussed extensively by Bertsimas and others [25].

Our model thus finds the optimal site locations (**Z**) and travel pairings (**X**) by maximizing the following objective function, subject to constraints.

$$\text{maximize} \left(\sum_{j=1}^{J} \boldsymbol{w_R z_j F_j} \right) + \boldsymbol{w_{\xi_1} \xi_1} + \boldsymbol{w_{\xi_2} \xi_2} + \boldsymbol{w_{\xi_3} \xi_3}$$

Subject to:

$\boldsymbol{x_{ij}} = 0 \text{ if } \boldsymbol{A_j} < \boldsymbol{A_{min}}$	Sites are built if there is enough available land	(1)
$\boldsymbol{x_{ij}} = 0 \text{ if } \boldsymbol{d_{ij}} > \boldsymbol{D_{max}}$	Sites are built if the distance from home to site is below a set threshold	(2)
$\boldsymbol{z_j} = 1 \forall \boldsymbol{j} \in \boldsymbol{L}$	We account for existing sites	(3)
$\sum_{j=1}^{J} \boldsymbol{z_j} = \boldsymbol{M}$	We build at most M sites	(4)
$\frac{1}{\lvert \boldsymbol{H_k} \rvert} \sum_{i \in \boldsymbol{H_k}}^{\boldsymbol{p_i^{(m)}}} \boldsymbol{x_{ij}} \geq \boldsymbol{\xi_m} \forall \boldsymbol{k}$	We find the best lower bound on the proportion of individuals in the mth demographic group for each group	(5)
$\boldsymbol{x_{ij}} \leq \boldsymbol{z_j}$	The non-zero columns of x map to the non-zero elements in z	(6)
$\sum_{j=1}^{J} \boldsymbol{x_{ij}} \geq \boldsymbol{z_j}$		

Fig. 3 Interface that filters and selects sites based on variables and criteria chosen

Our framework for site selection combines the tools from integer programming and data science to help GCF in three ways. First, the use of an optimization framework provides a quantitative assessment of the utility of the site as a function of available data. Second, a set of site summary statistics can be made readily available for a given optimal solution. Third, GCF members can tune the model to reflect their beliefs using hyperparameters. We developed the framework as a linear model to increase its interpretability for non-technical audiences. As GCF members may not be familiar with mathematical modeling, interpretability is key; the simple linear formulation allows for effectively communicating how data are incorporated into the model and how hyperparameter values can affect the cost function (Fig. 3).

5 Conclusions and Reflections

In this chapter, we presented the need to create technologies in context with communities through collaborative and reparative planning practices. We first presented the history of GIS and other computer mapping technologies to underscore the necessity of collaboration and transparency for contentious planning and design problems. Rational planning methods were insufficient to account for the essential connection between communities and the spaces they live in. We then highlighted our approach on using collaborative planning as the foundation for our technology. Traditions of participation and community design within the planning discipline can be traced back to the 1960s, but the scalability of those efforts has been limited and extensive community participation in software development has largely remained unexplored. We demonstrated the potential of this approach through our partnership with GCF. This case study offers both a framework for incorporating end user participation in technology development and concrete computational tools to facilitate community engagement in the built environment. Overall, we demonstrate how we can build technologies that flow from community values, focus on community knowledge, and center our technologies on people and outcomes, not on products.

5.1 *Ethical Technologies Through Collaborative Planning*

The approaches presented in this chapter extend best practices in participatory design to the design of software—itself a growing part of urban infrastructure—to exemplify how computing can help with building more resilient, just, and ethical spaces. We have shown how a participatory and community-based approach to technology development can result in long-term trusted partnerships, accessibility, and adoption of technologies by communities, and more scalable and impactful long-term impacts of technologies in community-based planning practices.

One thing to note about these interventions is the tension of competing narratives. The unclear narratives around data stewardship made it difficult to understand which data streams should be allowed for access. Technologies like motion sensing were deemed overly invasive, but use of technologies like Salesforce encouraged thinking about interactions with volunteers in a customer context, which seems at odds with GCF's charitable mission. Although it is not our role to challenge GCF's chosen data practices, moving forward with our partnership will require more careful analysis of these data practices, especially given current debates about the ethics of data collection. As this partnership has shown us, our path forward will only be possible through close and intentional engagement with the organizations and populations that will be affected by our work.

5.2 *Tech Education and Long-Term Robustness*

The history of technology has shown us that marginalized communities have been typically left behind; the rapid advancement in AI has exacerbated these concerns. Our proposed approaches for technology development directly address this concern by incorporating GCF members into the technology creation process. Both small-scale digitization approaches described offer opportunities for GCF corps members to get involved and learn about related technologies. The sensor module platform will serve as an introduction to electronics, firmware programming, and time series data analytics.

Further, the site allocation model can offer an introduction to the use of geospatial information to empower and inform decision-makers. By incorporating GCF corps members into the design of new technology, our approach serves as an education and outreach opportunity in an indirect manner. Additionally, offering these opportunities will likely increase interest in the GCF program, improving recruitment. Future work, currently in development, will incorporate teaching of technology development in outreach programs as a way of empowering underserved communities and providing tools for autonomous technology development.

These questions of long-term robustness are particularly important given GCF's mission on climate resilience. GCF takes large-scale concerns—ex. the unequal impact the climate crisis will have on traditionally marginalized populations—and addresses them on the local scale. GCF's holistic approach has been successful at addressing instances of historical and systemic spatial, environmental, and racial injustice through climate and green infrastructure jobs and community engagement. We seek to continue this "think global, act local" approach within the technology we develop ourselves, ensuring sustainability both technologically and environmentally.

Acknowledgements This work has been supported by the MIT Norman B. Leventhal Center for Advanced Urbanism through the 2022 Norman B. Leventhal City Prize and by the SERC Scholars Program.

References

1. Olascoaga CS (2021) Drawing participation: histories of geospatial computing, professional silos, and potentials for collaboration in planning and design. Thesis, Massachusetts Institute of Technology
2. Tomlinson RF (1988) The impact of the transition from analogue to digital cartographic representation. Am Cartogr 15(3):249–262. https://doi.org/10.1559/152304088783886937; Horwood E (1963) Using computer graphics in community renewal: computer methods of graphing, data positioning, and symbolic mapping. University of Washington, Seattle; Coppock JT, Rhind DW (1991) The history of GIS. In: Maguire DJ, Goodchild MF, Rhind DW (eds) Geographic information systems: principles and applications, vol 1. Longman Group UK Ltd, London; Foresman TW (1998) The history of geographic information systems: perspectives from the pioneers. Prentice Hall PTR
3. Fisher H (1966) Professor Edgar M. Horwood correspondence. September 27, 1966, Papers of Howard T. Fisher, ca. 1929–1974
4. de Monchaux N (2016) The map and the territory. In: Local code—3,659 proposals about data, design & the nature of cities. Princeton Architectural Press, Princeton, NJ, pp 154–177
5. Harvard University, Laboratory for Computer Graphics and Spatial Analysis, Fisher H (1975) SYMAP. In: Summary of research at the Laboratory for Computer Graphics and Spatial Analysis. Cambridge, MA
6. Computer map-making. In: Harvard alumni bulletin, March 5, 1966
7. Chrisman N (2006) Charting the unknown: how computer mapping at Harvard became GIS. ESRI Press, Redlands, CA
8. Fisher H (1965) Letter to Mr. Louis Winnick, Associate Director, Public Affairs, The Ford Foundation. January 14, 1965, Papers of Howard T. Fisher, ca. 1929–1974

9. Fisher H, Nash W (1965) Proposal for an educational and research grant relative to computer science and systems analysis applications to city planning. September 28, 1965, Papers of Howard T. Fisher, ca. 1929–1974
10. Fisher H (1973) Letter to Mrs. Jack Dangermond, Environmental Systems Research Institute. January 24, 1973, The Howard Fisher Paper Archive
11. Stadelmann M, Derrenbacher B (2019) Interview with Mirjam Stadelmann and Bill Derrenbacher at ESRI's Headquarters. November 22, 2019
12. Dangermond J, Freedman C (1986) Findings regarding a conceptual model of a municipal database and implications for software design. Geo-Processing 3:31–49
13. Morehouse S, Derrenbacher B (2019) Interview with Scott Morehouse and Bill Derrenbacher at ESRI's Headquarters. November 21, 2019
14. Aslıgül Göçmen Z, Ventura SJ (2010) Barriers to GIS use in planning. J Am Plan Assoc 76(2):172–183. https://doi.org/10.1080/01944360903585060
15. Arnstein SR (1969) A ladder of citizen participation. J Am Inst Plann 35(4):216–224. https://doi.org/10.1080/01944366908977225
16. Flyvbjerg B (2004) Phronetic planning research: theoretical and methodological reflections. Plan Theory Pract 5(3):283–306. https://doi.org/10.1080/1464935042000250195
17. Recent complaints have focused on "NYCHA's failure to provide decent, safe, and sanitary housing", and have made available local and federal funding for capital improvements. See Yalen RW et al (2018) United States of America v. New York City Housing Authority. Southern District of New York
18. Williams RA (2020) From racial to reparative planning: confronting the white side of planning. J Plan Educ Res. https://doi.org/10.1177/0739456X20946416
19. Williams
20. Arnstein A ladder of citizen participation
21. Yu C et al (2020) Plant spike: a low-cost, low-power beacon for smart city soil health monitoring. IEEE Internet Things J 7(9):9080–9090. https://doi.org/10.1109/JIOT.2020.3003479
22. PurpleAir. https://www2.purpleair.com/. Accessed 4 Apr 2023
23. US Environmental Protection Agency. Air quality system data, internet database. https://www.epa.gov/outdoor-air-quality-data. Accessed 4 Apr 2023
24. Urban Food Policy Institute (2019) Farms at NYCHA: final evaluation report. CUNY, New York, NY
25. Bertsimas D, Farias VF, Trichakis N (2011) The price of fairness. Oper Res 59(1):17–31. https://doi.org/10.1287/opre.1100.0865; Bertsimas D, Tsitsiklis J (1997) Introduction to linear optimization, 1st edn. Athena Scientific

Poor Data Quality in Sub-Saharan Africa and Implications on Ethical AI Development

Leonard Francis Vibbi

1 Introduction

Africa is experiencing a significant population growth, with its populace of 1.1 billion projected to reach 1.7 billion by the year 2030 [1]. The continent of Africa, encompassing 54 countries, exhibits vast geographical and cultural diversity. Despite its substantial population and diverse cultural landscape, the sub-Saharan Africa (SSA) region is currently striving to keep pace in the global endeavor to leverage big data for artificial intelligence (AI), a vital driver of innovation and economic advancement [2]. In light of this consideration, the assessment of SSA's utilization of big data in AI applications can be effectively conducted using a proxy measure that reflects the underrepresentation of SSA in the global adoption of big data for AI. As reported in the 2023 Artificial Intelligence Index Report, an evaluation based on peer-reviewed AI journals, which serves as a reliable indicator of AI research, reveals that SSA's contribution accounts for a mere 0.77%, representing the lowest proportion worldwide. The majority of publications in this field predominantly originate from regions such as East Asia and Pacific, Europe and Central Asia, and North America [3]. The challenges within Africa's data ecosystem contribute to the limited growth of AI based technology development at scale in the region, which can be attributed to a complex interplay of socio-technical factors. These factors encompass the presence of a digital divide, inadequate human capital or capacity, and other associated issues. The digital divide encompasses the discrepancy in access to technology and internet connectivity among different regions, resulting in unequal opportunities for data collection and utilization. Insufficient human capital or capacity refers to the scarcity of skilled professionals and resources necessary for effective data-driven research and AI development. These multifaceted factors collectively pose potential challenge to ethical AI development in Africa, emphasizing the necessity for targeted

L. F. Vibbi (✉)
Media Lab, SERC, Massachusetts Institute of Technology, Cambridge, MA, USA
e-mail: vibbi@mit.edu

S. Chiodo et al. (eds.), *Improving Technology Through Ethics*,
PoliMI SpringerBriefs, https://doi.org/10.1007/978-3-031-52962-7_7

initiatives aimed at addressing the existing challenges and cultivating an enabling environment for data-driven innovation.

2 Overview of Poor Data Quality in Sub-Saharan Africa

2.1 Insufficient Data Availability

SSA encounters notable obstacles in terms of data availability across multiple sectors, notably in the areas of health and the environment. Concerning the environment, specifically in relation to climate change, there exists a significant scarcity of data and limited access to such data. These challenges can be attributed to persistent issues of inadequate investment in climate operations in Africa, as well as the prevailing digital divide across the continent [4]. In the health sector there is critical limited data on the prevalence of diseases such HIV, obesity, and chronic obstructive pulmonary disease (COPD). A 2014 study on association between BMI and diabetes in Africa in SSA, was done with nearly 20 SSA countries having no data for diabetes. Similarly, there is limited data on the morbidity, mortality, and economic burden COPD [5, 6]. Moreover, the agriculture sector in Africa is significantly impacted by data quality issues. As per the FAO, only two out of the 44 sub-Saharan African countries assessed have high data collection standards, while 21 countries have low standards (FAO, 2008). This challenge is exacerbated in the poorest countries, where agriculture plays a critical role in livelihoods, resulting in inadequate data due to limited resources. Consequently, these countries struggle to allocate funds for improving data quality, making it challenging to generate sufficient survey or administrative data for informed policymaking. Regrettably, the financial constraints faced by these countries not only impede policy funding but also hinder their ability to access the necessary empirical evidence to support policymaking [7]. Factors such as inadequate infrastructure, digital divide, and limited resources for data collection, contribute to limited data availability in Africa. Only 28–33% of the population uses internet, meaning that around an estimated 871 million people are not realizing digital dividends. To this population, the power of information and communication technologies (ICT) to connect; to deliver goods and services in healthcare, education, finance, commerce, governance, and agriculture; to ease communication; and to share information cannot be actualized, and neither can their economies harness the power of technologies as a driver for economic growth and innovation [8, 9].

2.2 Data Incompleteness and Inconsistency

Data incompleteness and inconsistency contribute to poor data quality in SSA. Incomplete and inconsistent data hinder the accuracy and reliability of data thereby contributing to their poor-quality state. In SSA, several factors can be attributed to this such as the inadequacy of infrastructure and limited resources which became evident during the global pandemic. The Covid-19 pandemic revealed numerous African countries that faced challenges such as the absence of ICT infrastructure, inadequate network coverage, and restricted access to digital devices to meaningfully collect surveillance data to mitigate the Covid-19 infection. Additionally, the high costs associated with establishing ICT infrastructure and the lack of energy further hindered effective public health responses by limiting ICT interventions that rely heavily on access to data. This dearth of data availability obstructed efforts to address critical issues in sectors such as health, education, trade, industry, communication, banking, and labor, hampering the formulation of informed solutions [10]. Poor infrastructure, SSA faced challenges in generating accurate Covid-19 mortality data due to weak or lacking Civil Registration and Vital Statistics (CRVS) [11]. Insufficient data collection infrastructure in the majority of African nations has resulted in incomplete or absent data pertaining to previous years' mortality figures, crucial for assessing the extent of Covid-19-related deaths at a time of absence of accurate testing for Covid-19. As indicated by the United Nations, only eight countries, namely Algeria, Cape Verde, Djibouti, Egypt, Mauritius, Namibia, Seychelles, and South Africa, have managed to record over 75% of their respective mortalities. Notably, Ethiopia, a densely populated country and a prominent economy within Africa, possesses less than 2% coverage of its CRVS system [12].

3 Ethical Implications of Poor Data Quality on AI Development

The insufficiency of accessible data and the presence of incomplete and inconsistent data in SSA have profound ethical implications for the development of AI technology. Since AI development heavily relies on comprehensive training data, the limited availability of data, coupled with the incompleteness and inconsistency of existing data, inevitably leads to the creation of AI technologies that may perpetuate existing biases, unintentionally produce adverse effects, and impede transparency and accountability in their design and deployment. Although there is a scarcity of literature specifically addressing this issue in the context of SSA due to its distinctive social and cultural dynamics; there is substantial evidence suggesting that AI has the potential to reinforce prevailing biases across various demographics such as among women and people of color.

3.1 *Insufficiency of Accessible Data*

In order to understand the sources of biases in AI, it is essential to consider the socio-technical causes of bias, the manifestation of bias, and the definition of fairness (Fig. 1). Bias is injected into AI technologies when challenges arise during the data generation and collection process, including issues of inadequate data quality, ensuring data representativeness, incorporating sensitive features, and establishing an accurate fairness definition. These challenges significantly increase the risk of embedding biases specially given AI heavily relies on data sourced from human-generated content or collected through human-designed systems such as the web or internet of things (IoT). Consequently, biases inherent in human behavior through stereotypes and prejudice is encoded in the way data is collected and processed. Consequently, the deployment of biased AI systems in decision-making processes perpetuates discrimination and reinforces existing social inequalities [13].

A common example of how human bias is mirrored in AI technology bias can be the use of criminal risk assessment algorithms, which aim to evaluate the likelihood of a defendant reoffending by analyzing their profile. These tools provide a recidivism score, a single numerical estimate that judges use to make various decisions, including determining the appropriate rehabilitation services for individual defendants [14]. More empirically, a study on human versus AI bias in predicting recidivism conducted and published in [15] revealed that study participants demonstrated a false-positive rate of 37.1% for black defendants and 27.2% for white defendants, with a significant difference based on race (P = 0.027). The false-negative rate for black defendants was 29.2%, compared to 40.3% for white defendants, also showing a significant difference (P = 0.034). These findings mirror those of COMPAS, an AI predictive technology tool, which had a false-positive rate of 40.4% for black defendants and 25.4% for white defendants, with a significant difference (P = 0.002). COMPAS also exhibited a false-negative rate of 30.9% for black defendants and 47.9% for white defendants, with a significant difference (P = 0.003). Therefore, both the study participants and the COMPAS system displayed similar levels of unfairness towards black defendants, despite race not being explicitly specified (Fig. 2).

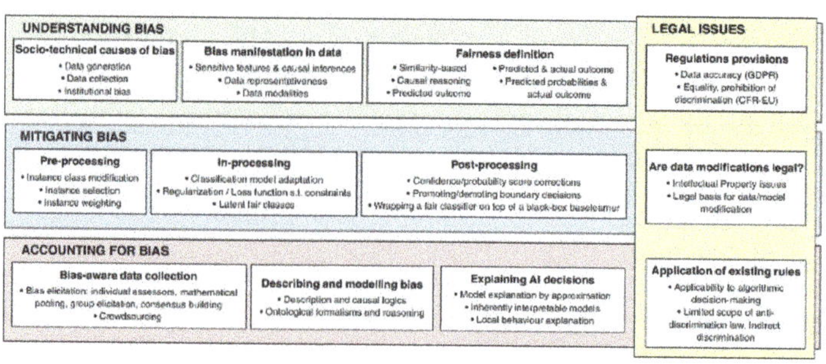

Fig. 1 Overview of topics related to bias in data-driven AI systems. Sourced from [13]

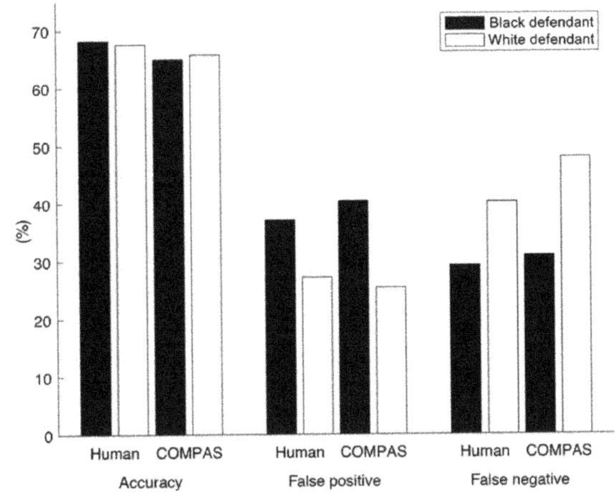

Fig. 2 Human (no-race condition) versus COMPAS algorithmic predictions. Sourced from [16]

3.2 *Limited Representativeness*

Furthermore, inaccurate or incomplete data can contribute to issues of limited representativeness of data particularly for marginalized groups. This may result in unintended consequences, as AI models trained on such data can generate misleading insights or recommendations, potentially harming individuals or communities. A common case attributed to the impact of incomplete data is reflected in the case of gender bias in AI algorithms. In particular gender ideologies encoded in text data used to train machine learning algorithms for AI technologies perpetuate common existing gender stereotypes. For example, [17] summarize five ways this is made evident. Three of these stereotypes embedded in AI as a result of inaccurate training data set is highlighted below.

1. *Naming*: Analysis of the terms "girl(s)" and "boy(s)" in British, American, and New Zealand English texts revealed that "girl" is three times more commonly used to refer to an adult. This usage portrays women as immature, innocent, youthful in appearance, subordinate in status, emotionally weak, or financially dependent. Additionally, using "girl" with occupations also diminishes the status of the job.
2. *Ordering*: The study on personal binomial order in the British National Corpus revealed gender as the primary factor influencing the order of naming pairs like 'man/woman', 'girl/boy', 'lady/gentleman', 'princess/prince', 'wife/husband', 'actress/actor', and 'he/she'.
3. *Bias Description*: Recent research has prioritized the investigation of gender bias in AI, specifically examining how word embedding algorithms acquire and propagate stereotypes. While researchers have predominantly relied on interpretation to identify stereotypical associations, a meticulous analysis of the British National Corpus revealed prominent gender stereotypes by scrutinizing the collocational patterns of terms related to men and women.

4 Addressing Poor Data Quality Challenges

The presence of insufficient data quality presents significant barriers to achieving transparency and accountability within AI systems. This deficiency impedes the ability to identify and rectify potential biases, errors, or discriminatory practices. These implications are particularly profound in the sub-Saharan African context, where the utilization of low-quality data for AI training can unintentionally perpetuate historical socio-cultural divisions, thereby perpetuating some of the most distressing historical narratives.

For instance, the impact of poor-quality data on the definition of citizenship in various African countries is noteworthy. Issues surrounding statelessness influenced by one's ethnic group can be reflected in determining who is considered a citizen. Consequently, AI technologies, when trained on such data, can reinforce notions such as digital identity exclusion at minimum and may extend to potential cases of digital-genocide for extreme cases, as marginalized groups and communities may effectively cease to exist due to their exclusion, lack of recognition as citizens and other attributable prejudicial treatment particularly targeted at race, ethnicity, religion, language, and culture [18]. Considering the aforementioned points, a relevant illustration of how bias can persist due to poor-quality data lies in the domain of criminal justice systems, as discussed in the preceding section. Another case study example is found in the loan or mortgage scoring algorithms, where minority groups often experience disadvantage due to the utilization of biased data during AI training. Specifically, lenders such as black people encounter significant challenges when assessing the risk of mortgage defaults, primarily attributable to limited data availability and information asymmetry. Consequently, this situation can result in either an overestimation or underestimation of the default risk, underscoring the importance of accounting for contextual factors and unfair biases in the assessment process [19]. Hence, examining the potential ethical issues such as biases stemming from AI trained on poor quality data in SSA can be illuminated by considering analogous cases in the criminal justice system, loan or mortgage scoring algorithms, as mentioned above. In sub-Sahara Africa bias encoded in data may perpetrate current states of political, social, and cultural discriminatory outcomes and suboptimal results. In order to address poor quality data and the related ethical issues the following recommendations are proffered.

5 Some Recommendations

5.1 Enhancing Data Collection Infrastructure

Investment in data collection infrastructure, such as improving internet connectivity and supporting digital literacy programs, can enhance data availability and quality. SSA has witnessed the most substantial worldwide growth in internet penetration, as the rate surged from under 1% in 2000 to 30% at present. Between 2019 and 2021, internet usage in Africa experienced a remarkable 23% leap. However, despite this impressive progress, a significant disparity persists, with more than 840 million individuals lacking access to dependable and affordable internet connectivity [20]. Addressing existing digital divide by increase in access to internet for unconnected communities can contribute to representative data that is complete and inclusive. In addition to internet access, data collection can also be improved by investing in digital literacy programs given the lag in digital literacy on the continent compared to other regions. Digital literacy can be best analyzed through the educational landscape in SSA which reveals a limited incorporation of "computer" skills within the school curriculum, encompassing only 50% of countries, in stark contrast to the global average of 85%. Consequently, an evident demand for digital skills among companies has emerged, substantiated by the findings in a report by World Bank, which also highlights a prevailing scarcity of proficient individuals in this domain [21]. The level of digital literacy in Africa is influenced by a multitude of factors. Of particular significance is the accessibility of cell phone network coverage and a dependable electricity grid, both of which are not universally accessible throughout the continent (Fig. 3). These prevailing conditions present formidable challenges to individuals seeking regular utilization of phones and computers [22]. The lack of adequate access to reliable internet and use of digital devices may naturally exclude data points from being accessible through online platforms for example. Hence, the provision of internet service for unconnected areas as well as ensuring access to regular supply of electricity will strengthen data collection infrastructure to become adequately accessible through use of digital devices.

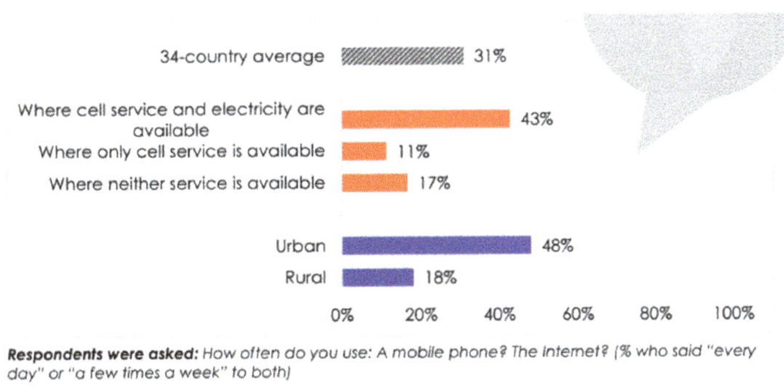

Fig. 3 Digital literacy index by neighborhood infrastructure and location, 34 countries, 2016/2018. Based on a survey conducted by Afrobarometer. Sourced from [19]

5.2 Promoting Data Diversity and Inclusivity

Encouraging the inclusion of diverse voices and perspectives in data collection processes can help address biases and ensure more representative datasets. This is highly crucial given the cultural and social diversity of SSA. SSA exhibits a notable array of diversity, characterized by distinct beliefs and traditional systems within its diverse communities, villages, and cities. The region holds ethnicity in high regard, deeply interwoven into the social fabric of most communities. This can be exemplified by the Hausa people, whose presence transcends national boundaries, spanning across Nigeria, Niger, Ghana, Chad, Cameroon, Cote d'Ivoire, and Sudan. The pervasive influence of ethnicity within SSA underscores its profound significance in promoting diversity and fostering inclusivity. Furthermore, ethnicity plays a crucial role as a fundamental constituent of demographic diversity, exerting direct influence over other dimensions of diversity. Recognizing and embracing one's ethnicity as an integral aspect of personal identity not only nurtures a sense of belonging but also contributes to the establishment of inclusive environments [23]. For data to be diverse and inclusive its collection and storage should mirror the portrayal of diversity—including the representation of all forms of visible and invisible differences including race, gender, ethnicity, disability, sexual orientation, education, class, rank, age, etc. [24]. Similar [23], data diversity and inclusivity should reflect the wide range of ethnicity in SSA which in turn may contribute to availability of diverse data training set that contributes to building and deployment of ethnically representative AI technologies. Which in effect can contribute to less bias and prevent prejudice especially for ethnically minority groups.

5.3 Establishing Data Quality Standards

Developing and implementing data quality standards, including data cleaning and validation processes, can help improve the reliability and accuracy of datasets. In the context of SSA, this is key especially given its cultural diversity reflected in ethnicity across geographical borders. Data quality standards depend on good data management reflecting criteria for accuracy, completeness, validity, consistency, uniqueness, timeliness, and fitness for purpose [25]. For SSA attention to detail and depth with what constitutes data quality should be noted and reflect community-value-driven systems. The presence of a diverse culture in SSA requires a detailed mapping of what data quality means. This then can be used to establish standards for data collection processes reflective of the context of cultures, ethnicity, age, disability, sexual orientation, education, class, rank, etc. [24].

5.4 *Collaborative Partnerships*

The promotion of collaborations among governments, private sector entities, and research institutions plays a vital role in facilitating the exchange of knowledge, enhancing capabilities, and advancing the development of ethically-driven data-based solutions tailored to the specific needs of SSA. Particularly, the cultivation of expertise in the realms of data ethics and ethical AI holds significant value. Noteworthy efforts by research think tanks like the African Center for Technology Studies (ACTS) actively implement initiatives such as AI for Development (AI4D) with the primary objective of fostering talent in the responsible domains of AI and machine learning within public universities across Africa. Supported by the International Development Research Centre and the Swedish International Development Cooperation Agency, this project spans a duration of three years and aims to address the growing demand for research and development in responsible AI and machine learning throughout the continent. The project specifically focuses on providing support to selected scholars engaged in PhD research in AI and machine learning within African universities, as well as early career academics seeking to enhance their capabilities in research and development within these fields. Significantly, the project places emphasis on research endeavors that center around responsible AI innovation for sustainable development, gender equity, and equitable distribution across regions with low-income countries [26]. The ACTS model effectively encompasses both public and private partnerships involving governance actors and development agencies. This collaborative framework serves as a crucial avenue for bridging the research gap in AI originating from Africa, while also facilitating the creation and sharing of knowledge that genuinely reflects the socio-technical context of SSA.

References

1. Spotlighting opportunities for business in Africa and strategies to succeed in the world's next big growth market. Brookings. https://www.brookings.edu/articles/spotlighting-opportunities-for-business-in-africa-and-strategies-to-succeed-in-the-worlds-next-big-growth-market/. Accessed 11 July 2023
2. Megwai S (2023) How AI is driving innovation and growth in Africa. Innovation Village | Technology, Product Reviews, Business, Mar 02, 2023. https://innovation-village.com/how-ai-is-driving-innovation-and-growth-in-africa/. Accessed 11 July 2023
3. HAI_AI-Index-Report_2023 (2).pdf
4. Dinku T (2019) Challenges with availability and quality of climate data in Africa. In: Extreme hydrology and climate variability. Elsevier, pp 71–80. https://doi.org/10.1016/B978-0-12-815998-9.00007-5
5. The burden of COPD in Africa: a literature review and prospective survey of the availability of spirometry for COPD diagnosis in Africa - Mehrotra - 2009 - Tropical Medicine & International Health - Wiley Online Library. https://onlinelibrary.wiley.com/doi/full/10.1111/j.1365-3156.2009.02308.x. Accessed 06 July 2023
6. Full text PDF. https://academic.oup.com/ije/article-pdf/46/5/1421/24575935/dyx078.pdf. Accessed 06 July 2023

7. Carletto C, Jolliffe D, Banerjee R (2016) From tragedy to renaissance: improving agricultural data for better policies. In: Statistical tragedy in Africa? Routledge
8. Africa Goes Digital – IMF F&D. https://www.imf.org/external/pubs/ft/fandd/2021/03/africas-digital-future-after-COVID19-duarte.htm. Accessed 06 July 2023
9. Munga J To close Africa's digital divide, policy must address the usage gap. Carnegie Endowment for International Peace. https://carnegieendowment.org/2022/04/26/to-close-africa-s-digital-divide-policy-must-address-usage-gap-pub-86959. Accessed 06 July 2023
10. Bakibinga-Gaswaga E, Bakibinga S, Bakibinga DBM, Bakibinga P (2020) Digital technologies in the COVID-19 responses in sub-Saharan Africa: policies, problems and promises. Pan Afr Med J 35(Suppl 2):38. https://doi.org/10.11604/pamj.supp.2020.35.2.23456
11. How chronic data gaps are hindering Africa's pandemic response. https://www.institute.global/insights/public-services/how-chronic-data-gaps-are-hindering-africas-pandemic-response. Accessed 01 May 2023
12. In Africa, lack of coronavirus data raises fears of 'silent epidemic'. Reuters, July 08, 2020. https://www.reuters.com/article/us-health-coronavirus-africa-data-insigh-idUSKBN24910L. Accessed 01 May 2023
13. Ntoutsi E et al (2020) Bias in data-driven artificial intelligence systems—an introductory survey. WIREs Data Min Knowl Discov 10(3):e1356. https://doi.org/10.1002/widm.1356
14. AI is sending people to jail—and getting it wrong. MIT Technology Review. https://www.technologyreview.com/2019/01/21/137783/algorithms-criminal-justice-ai/. Accessed 07 July 2023
15. Full text PDF. https://www.science.org/doi/pdf/10.1126/sciadv.aao5580. Accessed 07 July 2023
16. Dressel J, Farid H (2018) The accuracy, fairness, and limits of predicting recidivism. Sci Adv 4(1):eaao5580. https://doi.org/10.1126/sciadv.aao5580
17. Gender bias in artificial intelligence. In: Proceedings of the 1st international workshop on gender equality in software engineering. https://dl.acm.org/doi/abs/10.1145/3195570.3195580?casa_token=etlFbep3AHsAAAAA:akYMX4w5XijEU9aNZAUAZmUp5b9Ci0j1tGsO2qG5W3yUvIz9BxVAwwUfN_XHI9MiGyzVhpO17PmN3w. Accessed 07 July 2023
18. Aleinikoff TA, Klusmeyer D (2010) Citizenship today: global perspectives and practices. Brookings Institution Press
19. Lee MSA, Floridi L (2021) Algorithmic fairness in mortgage lending: from absolute conditions to relational trade-offs. Minds Mach 31(1):165–191. https://doi.org/10.1007/s11023-020-09529-4
20. The Internet Society pledges to expand Internet access in Africa. Internet Society. https://www.internetsociety.org/news/press-releases/2022/the-internet-society-pledges-to-expand-internet-access-in-africa/. Accessed 10 July 2023
21. Africa's future is bright—and digital, Oct 23, 2019. https://blogs.worldbank.org/digital-development/africas-future-bright-and-digital. Accessed 10 July 2023
22. Krönke M (2020) And the promise of e-learning
23. Appiah EK, Arko-Achemfuor A, Adeyeye OP (2018) Appreciation of diversity and inclusion in sub-Sahara Africa: the socioeconomic implications. Cogent Soc Sci 4(1):1521058. https://doi.org/10.1080/23311886.2018.1521058
24. O'Mara J, Richter A Global diversity and inclusion benchmarks: standards for organizations around the world
25. What is data quality? | IBM. https://www.ibm.com/topics/data-quality. Accessed 11 July 2023
26. User S Artificial Intelligence for Development (AI4D) Africa Scholarship Programme. https://www.acts-net.org/research/projects/artificial-intelligence-for-development-ai4d-africa. Accessed 11 July 2023

Social and Technological Innovation: Cross-Fertilization Needed

Federico Bartolomucci, Giorgia Trasciani, and Francesco Gerli

1 Technology for Good: Kept Premises and Let Downs

Contemporary innovation is nowadays characterized by technology and its pervasion in humans' interactions. Boundaries between the physical and the digital world have become ephemeral, with continuous overlaps between the two that have transformed our lives and created an inseparable bond between humans and technologies. Humans are nowadays leaving in what Floridi [1] defined as the new "onlife" space, a condition in which the boundaries between the online and offline dimensions break down, and almost every activity of our daily life is transformed and influenced by technologies. This revolution does not only affect how we buy groceries, pay taxes or read the news but also our personality, human-to-human interactions, and social behaviors. Brought to its extremes this digital pervasiveness in our lives may even exacerbate phenomena such as the hikikomori syndrome [2].

At the same time, this condition is mirrored and enhanced by an unstoppable transformation in many business sectors changed or totally disrupted by technological innovation. The effects of this revolution have been alternatively positive or negative depending on the involved actors' capacity to adapt their operations and business models. Emblematic in these terms has been the spread of Uber, which has disrupted the taxi industry, to the point in which entire States (e.g., Hungary, Denmark) have decided to ban it to preserve the existing equilibrium, while other countries have massively adopted it, forcing the taxi industry to adapt its model.

Technological development and digitalization are megatrends that, intrinsically, are neither good nor bad. However, their applications are not neutral [3]. As a matter of fact all technological innovation are social in its underlying processes, diffusion,

F. Bartolomucci (✉) · G. Trasciani · F. Gerli
Department of Management, Economics and Industrial Engineering, Politecnico di Milano, Milan, Italy
e-mail: federico.bartolomucci@polimi.it

S. Chiodo et al. (eds.), *Improving Technology Through Ethics*,
PoliMI SpringerBriefs, https://doi.org/10.1007/978-3-031-52962-7_8

and broader impact [4]. In other words, every innovation includes social and environmental impacts. Ignoring this insight, or reducing that relationship to one of economic exchange, increases the risks to ignore the role that technological development can play in furthering social inequality and environmental degradation. Actually, if some technologies could help in solving complex societal challenges, some may easily abuse of law enforcement, economic systems, and lead to exacerbation of gaps between tech controllers and users and underserved or just under- or mis-skilled populations.

Over the last decade, considerable efforts have been made worldwide to enact technology and digitalization as positive transformative forces in society with the promise to solve market inefficiencies and better meet people's needs as well as protect our planet. This approach defined Tech for Good [5] is related to the intentional design, development and use of digital technologies to address societal challenges. It encompasses directional and proactive management of technology to solve great societal problems, from poverty to ageing society, up to climate change. The technology for good scenario is characterized by a moral inherent connotation and judgement (see the "for good" label) which is ontologically related to the orientation of the technological process to a purposeful, societal objective.

Even in this paradigm, many adopted a techno-positivist approach, which often brought to overlook negative externalities created by tech-innovations diffusion. To name one, the sharing economy paradigm, which started with the promise to redistribute wealth and save resources, has often turned to have unintended consequences and generate negative externalities on levels not even considered at its debuts. This is the case for example of cities' central neighborhoods' gradual conversion into agglomerates of short-term rental apartments which drive rental prices up and make it difficult for locals to live in city centers. A phenomenon that in the long period may dismantle the net of intangible assets (e.g. neighborly relations, local commercial activities, formal and informal congregations) that constitute the essence of a community and that may be sustained and nurtured only by residents. This phenomenon of city centers' gentrification has been only partially addressed in big cities as New York, San Francisco or Amsterdam, where municipalities decided to limit the use of apps like Airbnb, while it is still creating massive problems in smaller cities, villages and islands that do not have the capacity or critical mass to limit their use. This example displays how even technologies that were born with a social purpose, if not properly regulated or managed may create negative externalities. This is particularly true when they enter their scaling phase and end up being denatured from their initial scope.

However, technological innovation is moving fast, and often regulations can be only reactive and discipline the uses of a certain technology ex-post. Coming to most recent times, Generative Artificial Intelligence systems have surprised the world for their widespread adoption and for reaching 100 million users in few weeks. Nonetheless their potential uses, risks connected, and impacts on people and businesses are still widely unknown. In Italy, during spring 2023 the use of ChatGPT was suspended for few weeks in name of potential privacy risks and violations of the European General Data Protection Regulation (GDPR). However, the initiative from the Italian Data Protection Authority was isolated, and risks to appear clunky and late, with respect to a technology that had already bewitched millions of users without encountering any resistance at that time.

In this chapter we put forward, therefore, the idea that, to avoid innovation's backlashes and strengthen our collective capacity to leverage technology for good, we should change the way we develop innovation and its drivers. We do so under the widely diffused assumption that how we develop it (i.e. the innovation process) has important implications on its output (i.e. the technological artifact) and thus by better managing the former, we could influence the latter.

We believe that a potential source of inspiration could be the social innovation field. Therefore, we call on innovation professionals to take on a contamination act between conventional technological innovation paradigms and few foundational concepts of social innovation. We consider two main acts of contamination essential to this scope: (i) rebalancing innovation drivers towards a more equal redistribution of its value, and (ii) reconsidering the roles of the various actors in the innovation process. We also identify two arenas where cross-contamination could occur. The first addresses the governance, the management and development of innovation processes, while the second addresses the competencies involved. Finally, we emphasize the role that Universities could play in nurturing this contamination through teaching, research, and public engagement.

2 Distilling Social Innovation Pillars

As is typical for the definition of complex phenomena, the meaning and principles of social innovation have changed over time and are contextually and historically constrained. Nonetheless, as a result of the pressing need to address significant societal issues and the failure of conventional market-based solutions to do so, policy interest in social innovation has increased substantially over the past few years, bringing to some definitory efforts. The Bureau of European Policy Advisors defined social innovation as "innovations that are social in both their ends and means [...], new ideas that simultaneously meet social needs more effectively than alternatives and create new social relationships or collaborations. [...] Innovations that are not only beneficial to society but also increase its ability to act" [6]. This definition highlights two crucial aspects of social innovations: (i) the new social relationships they establish and integrate as part of the innovation process, and (ii) the social value created by its outcomes and its redistribution.

With reference to the former, social innovation implies indeed a change in social relations, involving new methods of participatively doing, organizing, structuring and knowing [7]. It does so at various levels, including individual, organizational, and systems level, with the last one referring to institutions, social structures and physical infrastructures [8]. Social innovation is therefore conceptualized as the instrument or the set of instruments that contributes to reshaping society as a more participative arena where people are empowered to foster innovative solutions to meet their own needs and those of others differently than previous streamlined solutions. In this respect significant changes involve not just how we use technology, but also changes in markets, regulation, culture, industrial networks, and infrastructure.

The second component relates to novel methods of social value production and redistribution. We frequently assume that private companies create technological value. However, their capacity to extrapolate value from innovative technologies frequently stems from a combination of private and public investments, as well as an institution-shaping activity [9], which fosters business growth and development. However, historically, when technological advancements have occurred and benefited specific companies, they have primarily produced value for shareholders while ignoring other stakeholders. Technological innovation is unquestionably a generator of economic growth; nevertheless, if unregulated, the value it creates may be inhomogeneous and leave disadvantaged groups behind [10]. By bringing together capitals and knowledge in a few regions and hands, technologies pose the danger of aggravating social and economic inequalities, widening the divide between social classes, and even exacerbating territorial disparities. At the same time, they represent an opportunity to redistribute power, increase user control, and re-distribute wealth concentration while fostering social development. This, for instance, has been the premise and promise of distributed ledgers technologies, which were created to redistribute control over financial transactions and improve transparency and user control, but have instead become fertile ground for speculative operations and financial backlashes.

The social innovation perspective calls for the public actor, on behalf of the community it represents, to fully embrace its role and be involved in the redistribution of financial and social benefits. However, while policymakers continue to be concerned about supporting the introduction and dissemination of new technologies, very little is discussed and done regarding the reappropriation and redistribution of value.

On the other hand, if historically social innovation as a concept has run complementarily to the "hard" side of technological innovation, the complexity of societal grand challenges calls to combine the two perspectives in a holistic ones [11]. To unleash the transformative power of technology to create distributed social value, social innovators must confront and intertwine their practices and processes with conventional technological innovators, for the benefit of the economies and societies they act upon and vice versa.

3 Rethinking Tech Innovation as Social-Tech Innovation

In light of the two aforementioned factors, we propose revisiting our conception of technological innovation and redefining it as Social-Tech Innovation according to the following features.

Science and technology seek to provide answers and solve problems, but which questions are posed and whose problems are solved are contingent on whose voice prevails. Adoption of new inclusive technological innovation processes, fundamentally based on participation and co-creation of interventions, where the role of various actors is valued, could reconnect technological innovation to its role of creating shared value for the greater community and reposition it within the current social-ecological dynamics.

In this regard, particularly valuable is the role that the State and third-sector organizations might play.

The State and other public institutions have historically been mandated with resolving market and technological innovation inefficiencies, i.e., problems caused by technological innovation and free-market logics, through public spending or regulations. Historically, the State has fulfilled this function by de-risking activities, relieving businesses of systemic risks, and promoting technological diffusion through public infrastructure development. As Mazzucato [12] correctly notes, public funding has contributed in this way to the development of a number of tech giants, representing the frontier of tech innovation such as Facebook, Amazon, Apple, Netflix, and Google. However, this value, provided by the State and, by extension, by the community that it represents, is frequently overlooked.

Being among those who contribute to innovation, the State should claim and be recognized for its role of representing the community's interests and preventing rather than repairing potential social, environmental, and ethical technological harms.

This role will only be acknowledged and operationalized when public actors will be pro-actively involved, or better yet, when they will choose to be involved by promoting initiatives aimed at molding innovation and adopting an entrepreneurial mindset. In fact, although we tend to believe the government has little control over innovation, technological change is not random and does not occur in a vacuum.[1] Today, governments play a significant role in shaping the technological landscape, patent rules, and the definition and implementation of policies through R&D funding and education subsidies.

The third sector instead, comprised of not-for-profit and hybrid organizations, has historically occupied spaces where normal market logics make it inconvenient to operate, such as social assistance and rehabilitation services. Consequently, not-for-profit organizations are frequently the nearest to the most fragile and vulnerable segments of the population. In addition to relying more on technological innovation to solve their problems, these individuals are also the most susceptible to the unintended consequences of technological advancement. These segments of the population, including the elderly, the ill, and the young, require an intermediary that could both advocate for them and bring their cases to the attention of innovators, and, on the other end of the innovation process, assist them in avoiding risks associated with excessive or inappropriate exposure to technological innovation. In this regard, the not-for-profit sector is the ideal candidate and must be included as an intermediary actor between those in need and technologies throughout the various phases of the innovation process.

[1] To read more on this see https://www.project-syndicate.org/commentary/shaping-technological-innovation-to-serve-society-by-dani-rodrik-2020-03.

Involving both public and third sectors in innovation processes has the potential to level the playing field in terms of what types of innovations are created, how they are developed, and who benefits from them. A broader coordination between the public, not-for-profit, and private sectors is necessary to prevent, rather than solve, problems and to shape markets collectively, rather than repair them. In a multi-actor perspective, all economic actors, including government agencies, corporations, small businesses, third sector organizations, and citizen groups, should be empowered to collaborate [13]. This action of contamination and participation can activate collective intelligence in service of shared objectives and improve the equitable distribution of benefits among all stakeholders.

Involving the community's representatives and bringing the community's interests to the attention of technological innovators could bring about a positive revolution in the way technology is used to facilitate civic engagement and mutually empower and assist citizens and government, thereby exceeding citizens' expectations in terms of speed, quality, and accessibility.

4 Cross-Fertilization Arenas

As mentioned previously, technological innovation is directly entwined with socio-technical transitions and the social cultural context in which it occurs and is subsequently adopted. In this light, it calls for a reconsideration of the main reference processes of technology management, as well as innovation systems and ecosystems to allow for a greater influence between the affected societal and technological components and actors.

Technological innovation cannot be limited to the technological product and its market value. Technological artifacts can share the characteristics of social innovation when they proactively or intentionally aim to solve a societal problem and limit their externalities, when they are accountable for the societal impacts they generate, and when they actively engage with the demand side, i.e. the carriers of societal needs, during the solutions development process.

To put it into practice, we call for an in-depth cross-fertilization between the social innovation factors outlined in previous sections and the processes that contribute to technological development. However, it is not easy to determine in which spaces this desirable cross-fertilization could occur. Therefore, we identify two complementary and prospective cross-fertilization arenas, respectively referring to the inter-organizational dimension and the intra-organizational once.

(1) The first "arena" of cross-fertilization relates primarily to the innovation process, specifically the phases of design, prototyping, and diffusion of technological artifacts.

We envision a new role for target groups or beneficiaries, who should transition from being passive consumers to prosumers. Candidate users/beneficiaries should be involved in the collective identification of societal needs from the earliest stages of technology design, prototyping, and dissemination. Involvement of the target group should be aimed not only to comprehend their requirements, but also to anticipate their use of technology and its potential positive and negative consequences.

Involving either the beneficiaries themselves or their representatives, in the event that the former do not have the possibility to participate in the discussion effectively (e.g., people afflicted by a serious disease), may aid in anticipating potential technology misuse and incorporating countermeasures into its design.

Participation of user communities in the design and preliminary testing phases enables the incorporation of citizens' interests, social and cultural heritage into the novel technological artefact through endogenization. Using living lab scenarios to test the technology beyond its anticipated and mainstream application areas may disclose risks associated with its diffusion that are beyond what developers can see and foresee.

By virtue of this collective design and "re-design" of technological applications, "socially developed" technologies could provide a solution with a deeper level of appropriateness [14, 15], transforming solutions towards a "context-based" paradigm, fitting the specifics of the socio-cultural context on the demand side, in a more people-centred and societally inclusive manner. This shift may increase the likelihood of acceptance, reduce rejection risks, and eliminate negative externalities.

To achieve this, technological companies must develop new communication, co-creation, and co-design capabilities to establish a "societal challenge-oriented" dialogue and interaction with new types of actors (public actors, civil society, social enterprises, etc.), enabling a more reflexive and context-oriented approach to innovation. Innovation required to solve grand-societal challenges is not "innovation as traditionally studied and stimulated with a bias towards supply push" [16, p. 450]. On the contrary, it possesses a holistic and multidimensional nature, relying on the concept of "epistemic boundary spanning" to encompass a comprehensive understanding of innovation that interconnects technological and societal dimensions. This type of innovation emphasizes the synthesis and integration of diverse elements, including traditional innovation practices. It involves the methodical compilation and rearrangement of diverse components within a dynamic sociotechnical framework. Doing this, it also promotes novel forms of collaboration among previously marginalized actors in technological innovation processes in "creative corporativism" perspective [16].

In line with this view on innovation, the Italian project Get It Twice, initiated by a number of private foundations and the Politecnico di Milano, tested an empowerment program for the orientation of innovative solutions designed to address local societal challenges in the field of healthcare and welfare services.

The initiative relied on a phase of needs identification (Call for needs) that was developed with the participation of healthcare and welfare ecosystem actors. The initiative also includes two integrated calls: the Call for Solution, which seeks to select up to five technological solutions that address the identified needs, and the Call for Matching, which seeks to identify entrepreneurs and organizations already operating in the sector that were capable of testing and co-developing start-up models based on the identified technologies.

This was accomplished by structuring a multi-step innovation process that began with the introduction of a call for needs, inviting organisations within the social economy to convey their technical innovation needs. The call for need consisted of a survey and two focus groups with prominent experts and practitioners of the welfare and health sector. The needs emerged were then disseminated to the Politecnico di Milano's research community with an invitation to propose existing or new solutions. Once the most prominent solutions had been chosen, researchers were given the opportunity to begin a dialogue with the organisations that had expressed the need their innovation was addressing in order to jointly refine the technological solutions and develop a pilot project.

In this regard, the Get It Twice process is an illustration of a technological innovation and transfer process that was able to contaminate the typical innovation and technology transfer process with two elements drawn from the social innovation context, namely the identification of societal needs and the participation of local civil society organizations operating within the context.

(2) The second prospective arena of cross-fertilization relates to the intra-organizational competencies and expertise that contribute to the technological innovation process.

In a globalised world, a technology promoted by a single company may reach and affect millions of individuals within a few weeks. A small inconsistency or ethical lapse in its application may be promptly used against or inadvertently impact thousands of people. The extremely rapid rate of modern technology diffusion necessitates the introduction of ethically and socially sound technologies that have already been tested for potential misuses. Private actors involved in the technological innovation process must address the fact that development of innovation is not only about optimising its functionalities and user's fit, but also about minimising the related ethical and societal risks while maximising technology's positive social and environmental impacts (among others).

In this regards, reconsidering the competencies that enter the tech innovation process is a means to avoid and anticipate these risks. In other words, technical and managerial skills are insufficient to deal with the societal and ethical implications of technologies. Tech companies need to develop new competencies and a new vocabulary and grammar for innovation development.

The incorporation of humanistic competencies, including philosophy, sociology, and anthropology knowledge, should become a requirement for tech developers in order to enhance their capacity to incorporate a sociotechnical perspective into their logics.

However, this act of contaminating competences to prevent risks arising from the exploitation and diffusion of technology is not risk-free and requires companies to commit to it at strategic level.

Between December 2020 and February 2021, Google replaced two of its Ethical AI team's most prominent researchers. One of them was Timnit Gebru, a leading researcher whose work at Microsoft, developed in collaboration with Joy Buolamwini of the MIT Medial Lab, pioneered research on the risk of bias in artificial intelligence [17]. The reason for her dismissal was a controversy surrounding a paper she was about to publish (and which was later published [18]), which was considered to be inappropriate from the company. The paper discusses numerous hazards associated with AI development and diffusion, including gender and racial discrimination, the spread of disinformation, and environmental costs.[2] The company refused to comment on the incident, but it is widely believed that they viewed the paper as a threat to the growth of their AI-related business. Approximately two years later, in the spring of 2023, the introduction of generative AI tools has sparked several alarming declarations and statements, including some from prominent AI developers. In response, the CEOs of the four largest AI-developing companies, Alphabet Inc's Google, Microsoft, OpenAI and Anthropic, were invited to the White House for a "frank discussion" in which they were asked to provide evidence that they are meeting the US President's expectations to "make sure their products are safe before making them available to the public".[3] Meanwhile EU and USA strengthened their cooperation to regulate AI development and create a shared common of conduct.

These various facts demonstrate how the cross-contamination of competencies and the incorporation of humanistic figures in tech companies could be advantageous for identifying and most of all anticipating technological development-related risks. Moreover, it illustrates how the management of these hazards is frequently in competition with other types of interests, primarily commercial ones. In this regard, there are two potential solutions: regulatory intervention by the State, and the prevention of such risks and the ability to deal with them ex-ante, while it is still feasible for organisations to do so without incurring excessive financial losses.

Only through competencies contamination technology companies could anticipate these risks and operationalize their triple objective of profit maximisation, risk avoidance, and positive impact generation. The expansion of their mandate necessitates a new and revised understanding of the role of innovation and the value of interdisciplinarity in the domain of technological innovation.

5 The Role of Universities

We have not yet discussed the role of the university system, which, through its three missions (e.g. research, teaching and public engagement), may play a significant role in promoting the cross-fertilization of traditional technological and social innovation. To begin with, we think scientific research communities operating in the development of hard technologies and STEM fields should cross-fertilize with the social innovation research community and vice versa, opening novel and common research agendas and enriching the scientific language, the methodologies, and the themes explored by each community.

---

[2] For more info: https://www.technologyreview.com/2020/12/04/1013294/google-ai-ethics-research-paper-forced-out-timnit-gebru/

[3] For more info see https://www.nytimes.com/2023/05/04/technology/us-ai-research-regulation.html.

For instance, each year, the School of Management of Politecnico di Milano organizes through its Sustainability Committee several interactions between its researchers, students, and representatives of organizations in the social economy sector. This ongoing interaction includes, among other activities, the Lab on Sustainable and Social Innovation, taught by Professor Irene Bengo, which engages students in the development of real-world case-based projects, such as the creation of sustainable business models for socially oriented tech enterprises. During the lab students are asked to interact with the organizations and develop an output responding to their requests. An additional important activity is the annual meeting between researchers and social economy actors, which aims to foster the emergence of research lines of common interest and the conception of activities that could bring the research community closer to the sector's needs. The 2022 meeting, held in November 2022, included a workshop aimed at matching the needs of the academic community with those of the not-for-profit sector. The workshop led to the emergence of three primary research areas: (i) impact measurement and the creation of accessible training occasions on the topic, (ii) change management strategies and models to foster the integration of advanced technologies into social economies' actors operational models, (iii) the creation of innovative platforms to foster the dialogue and integration of both actors operating in the same sector and actors operating in different sectors to contaminate their knowledges and spur innovation.

In addition, fostering innovative educational programs and research-based teaching is essential to enable the flourishing integration of social sciences, anthropology, philosophy, and design into the teaching curriculum of technical universities. This expanded approach aims to provide a holistic perspective on technologies, encompassing not only scientific and engineering subjects, but also these complementary disciplines.

Engineering practices inherently incorporate both technical and societal aspects throughout their various phases. Therefore, it is crucial to adopt a more comprehensive approach to engineering education that acknowledges the sociotechnical nature of these processes. This approach should consider the philosophical, social, cultural, and historical implications associated with technological choices.

By revising and broadening the teaching offer of technical universities to include social sciences, anthropology, philosophy, and design, a more well-rounded understanding of technologies can be achieved. This integration will contribute to a holistic view of engineering, where the multifaceted nature of technological advancements and their impacts on society are adequately addressed.

This perspective underlies what American universities have done in recent years, contaminating technical aspects of their work with more social aspects. As an example, the Human-Computer Interaction Initiative at the MIT offers the course Interactive Data Visualization and Society. The course, co-taught by professors Arvind Satyanarayan (Computer Science) and Sarah. Williams (Technology and Urban Planning), seeks to challenge students from both Schools to employ interactive data visualization in the context of big societal challenges. The Spring 2023 course required students to use data visualization to illustrate the migration phenomenon and the factors that motivate people to migrate primarily via smugglers' routes from

Central to North America. The course was made possible and enriched by the participation of The World Food Programme, which provided the data and mentored the students on the difficulties and challenges of utilizing data in the context of migration. The course combines the acquisition of more technical aspects of data visualization and programming, with design and storytelling knowledge, and moments of discussion on the ethics of data visualization and its potential societal impacts. This contamination of perspectives gives the students not only the technical knowledge to visualize data, but also the intellectual tools to responsibly approach their data visualization effort and locate them into its broader context.

Universities, which include technical universities, should therefore engage in social innovation activities to a greater extent as part of their public engagement. These tasks should not be separated from technological research tasks. On the contrary, they are required to facilitate knowledge exchanges between universities and civil society organizations.

6 Conclusions

In a world confronting increasing and pressing societal challenges, technological innovation has already assumed a novel role in augmenting the human capacity to face them. However, technological development always involves varying degrees of ethical, social, and environmental risk. Even when innovations were aiming to produce positive effects, their expansion beyond what was initially anticipated triggered viscous circles that could potentially harm thousands of people.

In this paper, we put forward a strategic suggestion for technological innovation that aims to minimize the negative externalities of tech development while maximizing its social and environmental benefits. In this regard, the processes, resources, and competencies derived from the field of social innovation can contribute to informing and shaping the various phases of technological development, thereby making technologies more acceptable, appropriate, and human-centered and reducing the societal risks associated with the diffusion of technologies with negative unintended consequences.

We believe the success of this cross-fertilization perspective is contingent on our ability to (i) empower and integrate "socially oriented" intentional actors, such as public and third sector organizations, into innovation processes, and (ii) create new logics of technological innovation value redistribution. For these paradigm shifts to occur, we need to introduce adjustments in the way we design, implement, and bring technology to market. To do so we have identified and discussed two arenas where cross-fertilization can occur: (i) the innovation process in its distinct phases and (ii) the competencies that enter in this process.

The evolution of these two arenas towards a cross-fertilized perspective opens a series of implications for a wide range of heterogeneous actors involved in innovation development processes, each of them playing a different role.

Tech-companies may participate in this integration shift by gradually modifying hiring requirements, contributing privately to the development of cross-fertilization spaces as company living labs, and fostering new partnerships with civil society organizations and additional social innovation actors. Social scientists have indeed proved to improve the capacity of tech firms to know and enter new markets. They are also trained to critically examine the societal impact of technology, considering ethical implications and potential risks. They can help tech firms identify and address ethical concerns, ensuring that products and services are developed and deployed responsibly.

The public sector may also contribute to the governance of these novel multistakeholder and cross-fertilized programs through the implementation of appropriate and genuinely transformative innovation policies [19] and trough the adoption of a more pro-active innovation approach to shape innovation directionality towards societal challenges.

The not-for-profit sector may work as an intermediary between tech companies, citizens and beneficiaries, giving voice to the latter's needs and helping companies to adapt their innovations accordingly.

Universities must review their educational, research and public engagement towards a more holistic approach, encompassing a socio-technical view of technological innovation.

Collectively, the various organizations and profiles should be provided with a common and integrated education in technological, social, and humanistic disciplines so that they can govern the pattern of technological development towards the resolution of existing societal challenges. Ecosystems and innovation processes nurtured by such exchange and diversity have the potential to gradually abandon technological determinism for a more social constructivist perspective suited to counteract the unintended consequences of technologies and their management.

References

1. Floridi L (2014) The fourth revolution: how the infosphere is reshaping human reality. Book Oxford University Press, UK
2. Sales-Filho GS, Bandeira ID, Argollo N, Lucena RM (2023) Hikikomori syndrome and digital technologies: a systematic review. Harv Rev Psychiatry 50–59. https://doi.org/10.1097/HRP.0000000000000362
3. Kranzberg M (1986) Technology and history: "Kranzberg's Laws." Johns Hopkins Univ Press Soc Hist Technol 27:544–560
4. Moulaert F, Mehmood A, MacCallum D, Leubolt B (2017) Social innovation as a trigger for transformations: the role of research
5. NESTA (2019) Tecnologie emergenti per lo sviluppo sostenibile Ringraziamenti
6. Bureau of European Policy Advisors (BEPA) (2010) Empowering people, driving change: social innovation in the European Union. Eur Comm
7. Haxeltine A, Avelino F, Pel B et al (2016) A framework for transformative social innovation. TRANSIT Work Pap 5(5):1–33
8. Grin JR, Schot J (2010) Transitions to sustainable development: new directions in the study of long term transformative change. Routledge, New York, NY

9. Esposito G, Nelson T, Ferlie E, Crutzen N (2021) The institutional shaping of global megaprojects: the case of the Lyon-Turin high-speed railway. Int J Proj Manag 39:658–671. https://doi.org/10.1016/j.ijproman.2021.06.001
10. Zehavi A, Breznitz D (2017) Distribution sensitive innovation policies: conceptualization and empirical examples. Res Policy 46:327–336. https://doi.org/10.1016/j.respol.2016.11.007
11. Borrás S (2019) Holistic innovation policy: theoretical foundations, policy problems, and instrument choices / Susana Borrás and Charles Edquist
12. Mazzucato M (2018) Mission-oriented in the European Union
13. Avelino F, Wittmayer JM, Pel B et al (2019) Transformative social innovation and (dis)empowerment. Technol Forecast Soc Change 145:195–206. https://doi.org/10.1016/j.techfore.2017.05.002
14. Kaplinsky R (2011) Schumacher meets schumpeter: appropriate technology below the radar. Res Policy 40:193–203. https://doi.org/10.1016/j.respol.2010.10.003
15. Dekker R, Franco Contreras J, Meijer A (2020) The living lab as a methodology for public administration research: a systematic literature review of its applications in the social sciences. Int J Public Adm 43:1207–1217. https://doi.org/10.1080/01900692.2019.1668410
16. Kuhlmann S, Rip A (2018) Next-generation innovation policy and grand challenges. Sci Public Policy 45:448–454. https://doi.org/10.1093/SCIPOL/SCY011
17. Buolamwini J, Gebru T (2018) Gender shades: intersectional accuracy disparities in commercial gender classification. Proc Mach Learn Res 81:77–91
18. Bender EM, Gebru T, McMillan-Major A, Shmitchell S (2021) On the dangers of stochastic parrots: can language models be too big? In: FAccT 2021—proceedings of the 2021 ACM conference on fairness, accountability, and transparency, pp 610–623. https://doi.org/10.1145/3442188.3445922
19. Diercks G, Larsen H, Steward F (2019) Transformative innovation policy: addressing variety in an emerging policy paradigm. Res Policy 48:880–894. https://doi.org/10.1016/j.respol.2018.10.028

Index

S. Chiodo et al. (eds.), *Improving Technology Through Ethics*, PoliMI SpringerBriefs, https://doi.org/10.1007/978-3-031-52962-7

GPSR Compliance
The European Union's (EU) General Product Safety Regulation (GPSR) is a set of rules that requires consumer products to be safe and our obligations to ensure this.

If you have any concerns about our products, you can contact us on

ProductSafety@springernature.com

In case Publisher is established outside the EU, the EU authorized representative is:

Springer Nature Customer Service Center GmbH
Europaplatz 3
69115 Heidelberg, Germany

www.ingramcontent.com/pod-product-compliance
Ingram Content Group UK Ltd.
Pitfield, Milton Keynes, MK11 3LW, UK
UKHW020704041026
15590UKWH00001B/458
9783031529610